Torsion-Free Modules

Chicago Lectures in Mathematics

torsion-free modules

Eben Matlis

The University of Chicago Press
Chicago and London

Chicago Lectures in Mathematics Series
Irving Kaplansky, *Editor*

The Theory of Sheaves, by Richard G. Swan (1964)
Topics in Ring Theory, by I. N. Herstein (1969)
Fields and Rings, by Irving Kaplansky (1969; 2d ed. 1972)
Infinite Abelian Group Theory, by Phillip A. Griffith (1970)
Topics in Operator Theory, by Richard Beals (1971)
Lie Algebras and Locally Compact Groups, by Irving Kaplansky (1971)
Several Complex Variables, by Raghavan Narasimhan (1971)
Torsion-Free Modules, by Eben Matlis (1972)

The University of Chicago Press, Chicago 60637
The University of Chicago Press, Ltd., London

Published 1972
Printed in the United States of America

ISBN: 0-226-51073-5 (cloth); 0-226-51074-3 (paper)
Library of Congress Catalog Card Number: 72-95974

TO

MAXINE MATLIS

CONTENTS

INTRODUCTION

The subject of torsion-free modules over an arbitrary integral domain arises naturally as a generalization of torsion-free Abelian groups. Since our knowledge of these groups is quite limited, it is not surprising that we know very little about the more general situation. It is logical in beginning our study to restrict our attention to modules of finite rank. Since these modules are obviously direct sums of indecomposable ones, we are immediately confronted with the problem of finding all <u>indecomposable</u> torsion-free modules of finite rank.

Every torsion-free module of rank 1 is clearly indecomposable. An initial hypothesis, arising from a naive attempt to generalize the theory of finite-dimensional vector spaces, might be that rank 1 modules are the only indecomposable ones. Unfortunately, it can be shown that this hypothesis fails, even for Abelian groups. A more profitable approach is to turn the problem around and try to characterize the integral domains that do have this property. In the process we will shed some light on the nature of torsion-free modules in general.

We have defined an integral domain to have property D, or to be a D-ring, if every indecomposable torsion-free module does in fact have rank 1. In a series of papers exploring this property we have been able able to give a complete characterization of integrally closed D-rings.

1

The solution may be stated quite simply: an integrally closed domain is a D-ring if and only if it is the intersection of at most two maximal valuation rings (with the same quotient field).

The component parts of this solution are relatively inaccessible to the general reader, being spread over a number of different journals. Furthermore, the individual papers depend heavily upon a number of earlier, more general, papers. There is a need therefore to make available a coherent account of the theory of D-rings that is as self-contained as possible. This need has provided the principal motivation for writing these notes. Partly this motivation is aesthetic in that the theory is a pretty one, and its inherent beauty can best be seen if the arms, legs, and head of the statue are assembled and connected to the torso of the figure. For the other part, there is the hope that the methods and techniques displayed here will be found useful by other workers in the field in extending our knowledge of the properties of torsion-free modules.

Pursuing this purpose we have presented in the first eight chapters of these notes a general account of the theory of integral domains, re-quiring as a prerequisite only an elementary knowledge of commutative ring theory and homological algebra. Since the organizing principle has been to give just the necessary background for understanding the theory of D-rings, we have omitted many interesting topics; for instance the theory of almost maximal valuation rings, and also some of the relation-ships between torsion-free and torsion modules. These omissions have made it necessary to rework some of the proofs from their original published form. However, this has been a gain in that the arguments have usually become more transparent. The last seven chapters of the notes are devoted to the study of D-rings, using freely the results of the first eight chapters.

In chapters 1 and 2 we discuss the properties of cotorsion modules and completions as well as the relations between them. Without doubt there are no ideas in the general theory of integral domains which are more fundamental in nature than these. Completions play a role for torision-free modules analogous to that of completions for finitely generated modules over a Noetherian local ring. Of particular importance is the Duality Theorem which provides natural functorial equivalences between the categories of torsion-free, cotorsion modules and torsion, h-divisible modules. Furthermore, torsion-free modules are seen to be cotorsion if and only if they are complete; and thus completions can be described homologically in terms of the vanishing of the functor $\mathrm{Ext}_R^1(Q, \cdot)$.

In chapter 3 we prove the equivalence of a large number of definitions of an h-local ring, a very special kind of integral domain whose torsion and cotorsion properties are determined locally at each maximal ideal. It is an unexpected phenomenon that the theory of h-local rings provides a key link in the chain of solving the problem of rings with property D, as well as in problems concerning reflexive rings.

The next four chapters, 4 through 7, are concerned with the properties of reflexive rings. Apart from their own intrinsic interest, the theorems here will be vitally useful later in the chapters on D-rings. Reflexive rings are defined by the condition that every submodule of a

finitely generated torsion-free module is isomorphic to its double dual.
Thus we have another generalization of the theory of finite-dimensional
vector spaces, albeit of a different kind from that of D-rings. We may
therefore expect to discover at a later stage that there is a deep inter-
relationship between reflexive rings and D-rings. At this point we prove
that reflexive rings are h-local rings, a result that presages the role
that h-local rings will also play in the theory of D-rings.

Chapter 8 deals with many equivalent definitions of a maximal
valuation ring, and is the starting point for the whole theory of D-rings.
In fact the subject had its origin in the theorem (due to Kaplansky) that
a maximal valuation ring is a D-ring. One of the first serious con-
jectures was that there are no other D-rings. It was the failure of this
conjecture that led to the present theory. It remains true, however,
that a valuation ring is maximal if and only if it is a D-ring.

Property D is an hereditary property in the sense that if an integral domain has this property, then every one of its extension rings
(with the same quotient field) also has it. This hereditary quality is also
true for the defining properties of a maximal valuation ring. In some
metamathematical way hereditary properties that are shared by valuation rings usually state something about the integral closure of the ring.
But the reason for this can not be found merely in the fact that the integral closure is the intersection of valuation rings; for the intersection
of rings with a given property need not always have that property.

Another striking feature of the theory of D-rings is the recurrent
theme, like a leitmotif, of modules generated by two elements. This
theme is announced strongly in the chapters on reflexive rings. It is

picked up again in chapter 9 where we see that D-rings have the property that every finitely generated ideal can be generated by two elements (a result due to Bass). This property is hereditary in the sense described in the preceding paragraph; and we should therefore not be surprised to learn that the integral closure of a domain with this property is a Prüfer ring.

The two generator idea is so intriguing that we pause to characterize integral domains in which every ideal can be generated by two elements. We find that a domain has this property if and only if every finitely generated ring extension (with the same quotient field) is a reflexive ring and is Hausdorff in every ideal-adic topology. This is a kind of semi-hereditary property. The close connection of these ideas with the theory of D-rings is shown by the fact that a property equivalent to the two generator property is that the domain be Noetherian and that its localizations at each maximal ideal have a weaker, finitely generated version of property D.

With these theorems as clues, chapter 10 is devoted to finding all Noetherian D-rings. We define a ring of type II to be a complete, Noetherian, local domain such that every ideal can be generated by two elements. Such a domain necessarily has Krull dimension 1. We then discover that a Noetherian domain is a D-ring if and only if it is a ring of type II. This theorem destroys the hypothesis that a D-ring is integrally closed.

In chapter 11 we accomplish two things: with the aid of reflexive rings and of our two generator tools we show that the integral closure of a quasi-local D-ring is a maximal valuation ring; and guided by this result we construct an example of a non-Noetherian, quasi-local D-ring which is not integrally closed.

Chapters 12, 13, and 14 finally zero in on the problem of finding all integrally closed D-rings. The method of proof is not arbitrary, but proceeds step by step, forging a chain that leads to the ultimate solution; and, at least as far as this method of proof is concerned, every step is a necessary one.

First we prove by a direct construction of an indecomposable, torsion-free module of rank two that an h-local ring with more than two maximal ideals cannot be a D-ring. Then we prove that an h-local ring with more than one maximal ideal is a D-ring if and only if it is a ring of type I. (A ring of type I is defined to be the intersection of two maximal valuation rings that are so far apart that any domain which contains them both contains their common quotient field.)

At this point we need a new concept. We say that an integral domain has a remote quotient field if there is a torsion-free module of rank 1 that is not isomorphic to either the quotient field or to an ideal of the domain. Using a long and detailed argument that forms one of the major steps in the theory of D-rings we proceed to characterize rings of type I as precisely those D-rings that have a remote quotient field (a characterization obtained without assuming in advance that the ring is h-local). We deduce from this that rings of type I are the only noncomplete D-rings.

Our next step is to prove a theorem that substitutes for an inductive argument. This theorem states that an integrally closed domain is a D-ring if and only if both the factor ring and the localization with respect to a certain prime ideal are D-rings. From here we are within easy reach of the main theorem. After proving it we then proceed to construct examples of every possible kind of integrally closed D-ring.

The notes conclude with chapter 15 in which we show that Noetherian D-rings can be completely described as those D-rings which are Hausdorff in every ideal-adic topology.

Of course this summary of the chief results, and the means employed to obtain them, conceals the immense amount of creaking machinery that has to be built and lies about cluttering up the landscape. This machinery makes it difficult to see the path and follow the direction in which the argument is moving. At present there seems to be no way of avoiding this difficulty. The essense of the problem lies in the fact that in general there is no one canonical way of finding a direct sum decomposition of a given module. Homological algebra works up to a certain point in providing submodules which are always a direct summand in a given context. But usually the situation is more complex than this, and a module may split into a direct sum in a totally mysterious and noncanonical fashion. Nevertheless, homological algebra provides the skeleton on which the structure is built.

One of the distinctive features present in applying homological methods to classical module theory is the frequent recourse that must be made to a few duality-type identities. These identities condense a large amount of information about projective and injective resolutions into concise formulas. To make these notes more readable, three of these identities are listed here, together with their sources, and will be referred to in the notes by their designations as Theorem A1, Theorem A2,

or Theorem A3.   (The reference C. E. is to the book <u>Homological</u>
<u>Algebra</u> by H. Cartan and S. Eilenberg, published by Princeton Univer-
sity Press in 1956, item [5] in the bibliography.)

THEOREM A1:   (C.E., Chapter II, Proposition 5.2').   Let R and
S be rings, A a right R-module, C a right S-module, and B an R-S
bimodule with R operating on the left and S operating on the right.
Then we have a canonical isomorphism:

$$\text{Hom}_R(A, \text{Hom}_S(B, C)) \cong \text{Hom}_S(A \otimes_R B, C) \ .$$

THEOREM A2: (C.E., Chapter VI, Proposition 5.1).   With the
same data as Theorem A1, assume in addition that C is S-injective.
Then we have a canonical isomorphism:

$$\text{Ext}_R(A, \text{Hom}_S(B, C)) \cong \text{Hom}_S(\text{Tor}^R(A, B), C) \ .$$

THEOREM A3:   (C.E., Chapter VI, Proposition 4.1.3).   With
the same data as Theorem A1, assume that we have a ring homomorphism R → S, and that $\text{Tor}^R_n(A, S) = 0$ for all n > 0.   Then we
have a canonical isomorphism:

$$\text{Ext}_R(A, C) \cong \text{Ext}_S(A \otimes_R S, C) \ .$$

COTORSION MODULES

Our first task is to develop four fundamental exact sequences to which we will frequently refer in the text by their descriptions as (I), (II), (III), or (IV). Throughout these notes R will be an integral domain and we will assume that it is not a field. We will denote the quotient field of R by Q, and the R-module Q/R by K. Thus we have our first exact sequence:

(I) $$0 \to R \to Q \to K \to 0 .$$

An R-module A is said to be <u>torsion-free</u> , if given any nonzero element $r \in R$, the multiplication by r on A is a monomorphism. We will let $t(A)$ denote the torsion submodule of A; that is, $t(A) = \{x \in A \mid rx = 0 \text{ for some } r \in R, (r \neq 0)\}$. Then $A/t(A)$ is torsion-free, and A is torsion-free if and only if $t(A) = 0$. We will say that A is a <u>torsion module</u> if $A = t(A)$. A direct sum of modules is a torsion module if and only if every component of the sum is a torsion module.

Now Q is a flat R-module, and hence $\text{Tor}_1^R(Q, A) = 0$. Thus if we apply the functor $\cdot \otimes_R A$ to the exact sequence (I), we derive the second exact sequence:

(II) $$0 \to A/t(A) \to Q \otimes_R A \to K \otimes_R A \to 0 ,$$

and the isomorphism: $t(A) \cong \text{Tor}_1^R(K, A)$. It follows from (II) that A is a torsion module if and only if $Q \otimes_R A = 0$.

Dualizing the definition of torsion-free, we obtain the definition of divisible: B is said to be a divisible R-module if given any nonzero element $r \in R$, the multiplication by r on B is an epimorphism. An R-module B is said to be h-divisible, if it is an homomorphic image of an injective R-module. H-divisible implies divisible, and is a more useful concept. However, for torsion-free modules the definitions of divisible, h-divisible, and injective are all equivalent. If the only h-divisible submodule of B is 0, then B is said to be h-reduced. If we drop our h's, we still have valid definitions; and for torsion-free modules the two sets of definitions coincide.

We have a map $\text{Hom}_R(Q, B) \to B$ given by $f \to f(1)$ for $f \in \text{Hom}_R(Q, B)$. The image of this map is denoted by $h(B)$; and it is easily seen that $h(B)$ is the largest h-divisible submodule of B and contains every other h-divisible submodule. In general it is not true that $B/h(B)$ is h-reduced. Applying the functor $\text{Hom}_R(\cdot, B)$ to the exact sequence (I) we obtain the third exact sequence:

(III) $\qquad 0 \to \text{Hom}_R(K, B) \to \text{Hom}_R(Q, B) \to h(B) \to 0.$

Continuing this exact sequence we obtain the fourth:

(IV) $\qquad 0 \to B/h(B) \to \text{Ext}_R^1(K, B) \to \text{Ext}_R^1(Q, B) \to 0$.

We have already observed that A is a torsion module if and only if $Q \otimes_R A = 0$. Note that $\text{Tor}_1^R(Q, A) = 0$, since Q is a flat R-module. Dualizing this property we obtain the definition of cotorsion. An R-module C is said to be a cotorsion module if $\text{Hom}_R(Q, C) = 0$ and

$\text{Ext}_R^1(Q, C) = 0$. Thus, in particular, a cotorsion module is h-reduced. Clearly a direct product of modules is cotorsion if and only if each component of the product is cotorsion.

If C is a cotorsion R-module, then it follows from (III) and (IV) that $C \cong \text{Ext}_R^1(K, C)$. More generally, it can be shown that (just as $t(A)$ is the unique largest torsion submodule of an R-module A) if B is an h-reduced R-module, then $\text{Ext}_R^1(K, B)$ is the unique smallest cotorsion over-module containing B.

An R-module C is said to be <u>strongly cotorsion</u> if $\text{Ext}_R^n(Q, C) = 0$ for all $n \geq 0$. An R-module T is said to be <u>torsion of bounded order</u> if there exists a nonzero element $r \in R$ such that $rT = 0$. Since $\text{Ext}_R^n(Q, T)$ is then both torsion-free and torsion of bounded order, we must have $\text{Ext}_R^n(Q, T) = 0$ for all $n \geq 0$. Thus a torsion module of bounded order is a strongly cotorsion module.

The proof of the following theorem is immediate and is left to the reader.

THEOREM 1: Let $0 \to C' \to C \to C'' \to 0$ be an exact sequence of R-modules.

(1) If C' and C'' are cotorsion modules, then C is a cotorsion module.

(2) If C is a cotorsion module, then C' is a cotorsion module if and only if C'' is h-reduced.

(3) If C is cotorsion, and C' is strongly cotorsion, then C'' is cotorsion.

The following two theorems provide important ways of manu-facturing cotorsion modules.

THEOREM 2. If A is a torsion module, or if C is a cotorsion module, then $\operatorname{Hom}_R(A, C)$ is a cotorsion module.

Proof. By Theorem A1 we have $\operatorname{Hom}_R(Q, \operatorname{Hom}_R(A, C)) \cong \operatorname{Hom}_R(Q \otimes_R A, C)$. Thus if A is torsion, or if C is cotorsion, then $\operatorname{Hom}_R(A, C)$ is h-reduced. Now consider an exact sequence of R-modules:

$$0 \to C \to D \to B \to 0$$

where D is an injective R-module. Applying the functor $\operatorname{Hom}_R(A, \cdot)$ to this sequence we obtain an exact sequence

$$0 \to \operatorname{Hom}_R(A, C) \to \operatorname{Hom}_R(A, D) \to \operatorname{Hom}_R(A, B) .$$

Assume that A is a torsion module. Then by Theorem A2 we have $\operatorname{Ext}_R^n(Q, \operatorname{Hom}_R(A, D)) \cong \operatorname{Hom}_R(\operatorname{Tor}_n^R(Q, A), D) = 0$, and hence $\operatorname{Hom}_R(A,D)$ is a cotorsion module. Since $\operatorname{Hom}_R(A, B)$ is h-reduced, it follows from Theorem 1 that $\operatorname{Hom}_R(A, C)$ is a cotorsion module.

Now assume that C is a cotorsion module and consider an exact sequence of R-modules:

$$0 \to P \to F \to A \to 0$$

where F is a free R-module. Applying the functor $\operatorname{Hom}_R(\cdot, C)$ to this sequence we obtain an exact sequence:

$$0 \to \operatorname{Hom}_R(A, C) \to \operatorname{Hom}_R(F, C) \to \operatorname{Hom}_R(P, C).$$

Now $\operatorname{Hom}_R(F, C)$ is isomorphic to a direct product of copies of C, and hence it is a cotorsion module. Since $\operatorname{Hom}_R(P, C)$ is h-reduced by the first paragraph of the proof, it follows from Theorem 1 that $\operatorname{Hom}_R(A,C)$ is a cotorsion R-module.

THEOREM 3. Let A be an h-reduced R-module, and $\{C_\lambda\}$, $\lambda \in \Lambda$, a family of cotorsion submodules of A. Then $C = \bigcap_{\lambda \in \Lambda} C_\lambda$ is also a cotorsion submodule of A.

Proof. Let λ_0 be a fixed index in Λ, and define $D_\lambda = C_\lambda \cap C_{\lambda_0}$ for each $\lambda \in \Lambda$. We then have an exact sequence of the form:

$$0 \to D_\lambda \to C_\lambda \oplus C_{\lambda_0} \to C_\lambda + C_{\lambda_0} \to 0 .$$

Since $C_\lambda + C_{\lambda_0}$ is a submodule of A, it is h-reduced; and thus by Theorem 1, D_λ is a cotorsion module. It follows from another application of Theorem 1 that C_{λ_0}/D_λ is h-reduced.

Now $C = \bigcap_{\lambda \in \Lambda} D_\lambda$, and hence we have an exact sequence:

$$0 \to C \to C_{\lambda_0} \to \Pi\,(C_{\lambda_0}/D_\lambda).$$

Hence we can use Theorem 1 again to see that C is a cotorsion module.

The next two theorems are lemmas for the useful duality relationship expressed in Theorem 6.

THEOREM 4. Let B be a torsion R-module. Then the natural map $\psi: K \otimes_R \operatorname{Hom}_R(K, B) \to h(B)$ defined by $\psi(k \otimes f) = f(k)$ (for $k \in K$ and $f \in \operatorname{Hom}_R(K, B)$) is an isomorphism.

Proof. If we apply the functor $Q \otimes_R \cdot$ to exact sequence (III) we obtain an isomorphism : $Q \otimes_R \operatorname{Hom}_R(K, B) \to Q \otimes_R \operatorname{Hom}_R(Q, B)$, since $h(B)$ is a torsion module. There is also a canonical isomorphism $Q \otimes_R \operatorname{Hom}_R(Q, B) \to \operatorname{Hom}_R(Q, B)$ because $\operatorname{Hom}_R(Q, B)$ is a Q-module. If we compose these two isomorphisms we obtain an isomorphism: $\lambda: Q \otimes_R \operatorname{Hom}_R(K, B) \to \operatorname{Hom}_R(Q, B)$. There is also the canonical isomorphism $\eta: R \otimes_R \operatorname{Hom}_R(K, B) \to \operatorname{Hom}_R(K, B)$.

Applying the functor $\cdot \otimes_R \operatorname{Hom}_R(K, B)$ to the exact sequence (I), we obtain a diagram which is easily seen to be commutative:

$$
\begin{array}{ccccccccc}
0 & \to & R \otimes_R \operatorname{Hom}_R(K, B) & \to & Q \otimes_R \operatorname{Hom}_R(K, B) & \to & K \otimes_R \operatorname{Hom}_R(K, B) & \to & 0 \\
& & \downarrow{\eta} & & \downarrow{\lambda} & & \downarrow{\psi} & & \\
0 & \to & \operatorname{Hom}_R(K, B) & \to & \operatorname{Hom}_R(Q, B) & \to & h(B) & \to & 0 \ .
\end{array}
$$

The top row is exact, since $\operatorname{Hom}_R(K, B)$ is torsion-free; and the bottom row is exact by III. From this it follows immediately that ψ is an isomorphism.

THEOREM 5. Let A be a reduced, torsion-free R-module, and define a natural map $\phi: A \to \operatorname{Hom}_R(K, K \otimes_R A)$ by $\phi(x)(k) = k \otimes x$ for $x \in A$ and $k \in K$. Then we have an exact sequence:

$$0 \to A \xrightarrow{\phi} \operatorname{Hom}_R(K, K \otimes_R A) \to \operatorname{Ext}^1_R(Q, A) \to 0 \ .$$

Proof. Let $x \in \operatorname{Ker} \phi$, and let $r \in R$, $(r \neq 0)$. Then in $K \otimes_R A$ we have $(r^{-1} + R) \otimes x = 0$. Since A is torsion-free, it follows from the exact sequence (II) that there is an element $y \in A$ such that in $Q \otimes_R A$ we have $r^{-1} \otimes x = 1 \otimes y$. This implies that $x = ry$. Since A is reduced, we have $\bigcap_{r \neq 0 \in R} rA = 0$; and thus $x = 0$. Therefore, ϕ is a monomorphism.

Let $B = K \otimes_R A$ and $C = \operatorname{coker} \phi$. Then we have an exact sequence:

(a) $$0 \to A \xrightarrow{\phi} \operatorname{Hom}_R(K, B) \to C \to 0 \ .$$

Applying the functor $K \otimes_R \cdot$ to this sequence, and utilizing the facts that $\operatorname{Hom}_R(K, B)$ is torsion-free and that $t(C) \cong \operatorname{Tor}^R_1(K, C)$, we obtain an exact sequence:

(b) $0 \to t(C) \to B \xrightarrow{1 \otimes \phi} K \underset{R}{\otimes} \operatorname{Hom}_R(K, B) \to K \underset{R}{\otimes} C \to 0.$

It is easy to verify that if ψ is the map of Theorem 4, then $\psi(1 \otimes \phi)$ is the identity on B. Since B is a torsion, h-divisible module, ψ is an isomorphism by Theorem 4. Therefore $1 \otimes \phi$ is an isomorphism.

The exact sequence (b) now shows that $t(C) = 0$ and $K \underset{R}{\otimes} C = 0$. It follows from the exact sequence (II) for C that $C \cong Q \underset{R}{\otimes} C$. Hence we have $C \cong \operatorname{Hom}_R(Q, C)$. Since $\operatorname{Hom}_R(K, B)$ is a cotorsion module by Theorem 2, if we apply the functor $\operatorname{Hom}_R(Q, \cdot)$ to the exact sequence (a), we obtain the isomorphism: $\operatorname{Hom}_R(Q, C) \cong \operatorname{Ext}_R^1(Q, A)$. This proves the theorem.

THEOREM 6. (Duality): Let A and B be R-modules. Then
(1)   B is a torsion h-divisible module if and only if

   $B \cong K \underset{R}{\otimes} \operatorname{Hom}_R(K, B)$.

(2)   A is a torsion-free cotorsion module if and only if

   $A \cong \operatorname{Hom}_R(K, K \underset{R}{\otimes} A)$.

Proof. By exact sequence (II), $K \underset{R}{\otimes} \operatorname{Hom}_R(K, B)$ is a torsion, h-divisible module. Thus statement (1) follows from Theorem 4. Since K is divisible, $\operatorname{Hom}_R(K, K \underset{R}{\otimes} A)$ is a torsion-free module. It is a cotorsion module by Theorem 2. Thus statement (2) follows from Theorem 5.

The following theorem is a corollary of Theorem 6. This application of duality will play a vital role in the theory of D-rings for it will be the cornerstone of Theorem 89.

THEOREM 7.

(1) If B is a torsion, h-divisble module, then there is a one-to-one correspondence between direct sum decompositions of B and of $\text{Hom}_R(K, B)$.

(2) If A is a torsion-free, cotorsion module, then there is a one-to-one correspondence between direct sum decompositions of A and of $K \otimes_R A$.

Proof. This is an immediate consequence of Theorem 6.

2

COMPLETIONS

A topology called the <u>R-topology</u> is defined on an R-module A by letting the submodules of A of the form IA, where I is a non-zero ideal of R, be a subbase for the open neighborhoods of 0 in A. The same topology is obtained by letting the submodules of A of the form rA, where r is a non-zero element of R, be a subbase for the open neighborhoods of 0 in A. The R-topology on R makes R into a topological ring, and A into a (not necessarily Hausdorff) topological R-module. The R-topology is a uniform topology.

Let B be a submodule of A. In general the induced topology on B as a subspace of A is not the same as the R-topology on B. However, if B is pure in A (that is, if $rA \cap B = rB$ for every element $r \in R$), then the two topologies on B are the same. The closure of B in A in the R-topology is $\bigcap (B + rA)$, $(r \in R, r \neq 0)$.

Let $J(A) = \bigcap rA$, $(r \in R, r \neq 0)$; then $J(A)$ is the closure of 0 in A. A is Hausdorff if and only if $J(A) = 0$. Thus $A/J(A)$ is Hausdorff. It is easy to see that if A is torsion-free and reduced, then A is Hausdorff. In particular, R is Hausdorff.

The submodules $\{IA\}$, where I ranges over the non-zero ideals I of R, form a directed system under the containment relation. Thus we can form the inverse limit $\widetilde{A} = \varprojlim A/IA$. It is not hard to see that $\widetilde{A} = \varprojlim A/rA$, $(r \neq 0 \in R)$, and we will consistently use this representation of $\widetilde{A}$. Thus we view $\widetilde{A}$ as a submodule of the direct product $\Pi_A = \Pi A/rA$, $(r \in R, r \neq 0)$. We well-order the non-zero elements of R in a fixed arbitrary way, and we represent an element of Π_A in the form $<x_r + rA>$, where $x_r \in A$ and $r \in R$, $r \neq 0$. Then $<x_r + rA>$ is an element of $\widetilde{A}$ if and only if given any two non-zero elements s and t of R, we have $x_s - x_{st} \in sA$.

The module A/rA where $r \in R$, $r \neq 0$, is discrete in the R-topology, and hence Π_A is a product of discrete modules and has the product topology. Thus $\widetilde{A}$, as a subspace of Π_A, has a topology called the induced topology. We have a canonical R-homomorphism $j: A \to \widetilde{A}$ defined by $j(x) = <x + rA>$ for all $x \in A$. The kernel of j is $J(A)$.

We now let A be a torsion-free, reduced R-module. Then the kernel of j is 0, and we identify A with its image as a subspace of $\widetilde{A}$. It is easily seen that the R-topology on A and the topology induced on A as a subspace of $\widetilde{A}$ are the same, and that A is a dense subspace of $\widetilde{A}$ and a pure submodule of $\widetilde{A}$.

THEOREM 8. Let A be a torsion-free, reduced R-module endowed with the R-topology. Then:

(1) $\widetilde{A}$ is the completion of A.

(2) $\widetilde{A}/A$ is torsion-free and divisible.

(3) The induced topology coincides with the R-topology on $\widetilde{A}$. Thus $\widetilde{A}$ is complete in the R-topology, and A is complete if and only if $A \cong \widetilde{A}$.

Proof. That $\widetilde{A}$ is the completion of A is a standard piece of point-set topology and we will omit the details.

We will prove that $\widetilde{A}/A$ is torsion-free. Let $\widetilde{x} \in \widetilde{A}$, and suppose that $r\widetilde{x} = x$, where $x \in A$ and $r \in R$, $r \neq 0$. We write $\widetilde{x} = <x_s + sA>$, and then $rx_r - x \in rA$. Hence there exists $y \in A$ such that $x = ry$. Since $r(x_{sr} - y) \in rsA$ and A is torsion-free, it follows that $x_{sr} - y \in sA$ for every non-zero $s \in R$. Thus we have $\widetilde{x} = y \in A$.

We next prove that $\widetilde{A}/A$ is divisible. Let $\widetilde{x} \in \widetilde{A}$, and let r be a non-zero element of R. We write $\widetilde{x} = <x_s + sA>$; then for every non-zero $s \in R$ there exists an element $y_s \in A$ such that $x_{sr} - x_r = ry_s$. We define $\widetilde{y} = <y_s + sA>$, and it is not hard to show that since A is torsion-free, $\widetilde{y}$ is an element of $\widetilde{A}$. We define $x = <x_r + sA>$; then x is an element of A and $\widetilde{x} - r\widetilde{y} = x$. Hence $\widetilde{A}/A$ is divisible.

Finally, we prove that the induced topology and the R-topology coincide on $\widetilde{A}$. Let r be a non-zero element of R and define:
$$U_r = \{ \widetilde{x} = <x_s + sA> \in \widetilde{A} \mid x_r \in rA\}.$$
Since the sets of the form U_r form a subbase for the open neighborhoods of 0 for the induced topology on $\widetilde{A}$, it will be sufficient to prove that $U_r = r\widetilde{A}$.

It is clear that $r\widetilde{A} \subset U_r$. On the other hand let $\widetilde{x} = <x_s + sA> \in U_r$. Since $x_r \in rA$, there exists an element $y_s \in A$ such that $x_{sr} = ry_s$ for every non-zero $s \in R$. We define $\widetilde{y} = <y_s + sA>$, and since A is torsion-free it is not hard to show that $\widetilde{y}$ is an element of $\widetilde{A}$. We have $\widetilde{x} = r\widetilde{y}$, and this shows that $U_r \subset r\widetilde{A}$.

The remaining statements of the theorem are immediate consequences of what we have already proved.

The next theorem shows that completeness for reduced, torsion-free modules is a homological property and is described by the vanishing of $\text{Ext}_R^1(Q, \cdot)$. Thus we have available the machinery of homological algebra in settling topological questions, and conversely homological problems can be viewed in a fresh way from the standpoint of topology.

THEOREM 9. Let A be a reduced, torsion-free R-module. Then $\tilde{A}$ is a torsion-free, cotorsion R-module and $\tilde{A} \cong \text{Hom}_R(K, K \otimes_R A)$. Thus A is a cotorsion R-module if and only if A is complete in the R-topology.

Proof. By Theorem 8, $\tilde{A}/A$ is a torsion-free R-module, and thus $\tilde{A}$ is torsion-free. Since Π_A is a product of torsion modules of bounded order, it is a cotorsion R-module. For each non-zero $s \in R$ we define $\phi_s : \Pi_A \to \Pi_A$ as follows: If $z \in \Pi_A$, then $z = \langle z_r + rA \rangle$, and we define $\phi(z) = \langle (z_r - z_{rs}) + rA \rangle$. Clearly, ϕ_s is a well-defined R-homomorphism. Im ϕ_s is h-reduced since it is a submodule of the cotorsion module Π_A. Thus by Theorem 1, Ker ϕ_s is a cotorsion R-module. Since $\tilde{A} = \bigcap \text{Ker } \phi_s$ ($s \in R, s \neq 0$), we see by Theorem 3 that $\tilde{A}$ is a cotorsion R-module.

Since $\tilde{A}/A$ is torsion-free and divisible by Theorem 8, we see that $K \otimes_R A \cong K \otimes_R \tilde{A}$. Now $\tilde{A}$ is torsion-free and cotorsion, and thus by Theorem 6 we have $\tilde{A} \cong \text{Hom}_R(K, K \otimes_R \tilde{A})$. Therefore, $\tilde{A} \cong \text{Hom}_R(K, K \otimes_R A)$. Hence by Theorems 8 and 6, A is complete in the R-topology if and only if A is a cotorsion R-module.

The most important torsion-free R-module is R itself, and hence the most important complete R-module is the completion of R. In Theorem 10 we will see that the completion of R is a commutative ring; and in the following theorems we will examine the relationships of the ideal and module structures of the two rings.

THEOREM 10.   Let $H = \operatorname{Hom}_R(K, K)$; then H is isomorphis to $\widetilde{R}$, the completion of R in the R-topology. We have $R \subset H$ and $H/R \cong \operatorname{Ext}^1_R(Q, R)$. H is a commutative ring and a faithfully flat R-module.

Proof. We embed R in H by identifying elements of R with their multiplications on K. In the light of Theorems 5 and 9 we only need to prove the last sentence of the theorem. Let f be an element of H and r a non-zero element of R. Then $f(r^{-1} + R)$ is an element of K annihilated by r. Hence there exists $s \in R$ such that $f(r^{-1} + R) = s/r + R$. Thus if $x \in K$, we have $f(Rx) \subset Rx$.

Let g be another element of H, and suppose that $f(x) = ax$ and $g(x) = bx$, where $a, b \in R$. Then $f(g(x)) = f(bx) = bax = g(ax) = g(f(x))$. Hence $fg = gf$, and thus H is a commutative ring.

Since H/R is torsion-free and divisible, it is a flat R-module. It follows readily that H is a faithfully flat R-module.

THEOREM 11.

(1) Let T be a torsion R-module. Then the maps $\xi : T \to H \otimes_R T$ given by $\xi(x) = 1 \otimes x$, $(x \in T)$, is an isomorphism. Thus T has a unique structure as an H-module.

(2) Let C be a cotorsion R-module. Then the map $\eta : \operatorname{Hom}_R(H, C) \to C$ given by $\eta(f) = f(1)$, $(f \in \operatorname{Hom}_R(H, C))$, is an isomorphism. Thus C has a unique structure as an H-module.

Proof. (1) Because H/R is torsion-free and divisible we have isomorphisms: $T \cong R \otimes_R T \cong H \otimes_R T$ whose composite is ξ. Thus T has the structure of an H-module, extending that of R. This H-structure is unique. For suppose that we are given a map $\nu : H \otimes_R T \to T$

which puts an H-structure on T, extending that of R. Then for $x \in T$ we have $\nu(\xi(x)) = \nu(1 \otimes x) = x$, and thus $\nu = \xi^{-1}$.

(2) Because H/R is torsion-free and divisible we have isomorphisms: $\mathrm{Hom}_R(H, C) \cong \mathrm{Hom}_R(R, C) \cong C$ whose composite is η. Thus C has the structure of an H-module, extending that of R. This structure is unique. For suppose that we have an H-structure on C extending that of R, and denoted by $h \circ x$ for $h \in H$ and $x \in C$. We will denote the η-structure on C simply by hx. Fix $x \in C$ and define R-homomorphisms f and g from H to C by $f(h) = h \circ x$ and $g(h) = hx$. Then $\eta(f) = f(1) = x = g(1) = \eta(g)$. Since η is an isomorphism, $f = g$; and thus $h \circ x = hx$. Therefore, the η-structure on C is unique.

We need the following lemma for Theorem 13.

THEOREM 12. Let S be a commutative ring, I an ideal of S, and $f: S/I \to S/I$ an epimorphism. Then f is an isomorphism.

Proof. The kernel of f is J/I, where J is an ideal of S containing I. Then $S/J \cong (S/I)/(J/I) \cong S/I$. The annihilator of S/J is J, and the annihilator of S/I is I. Since S/J and S/I are isomorphic, we have $J = I$. Thus the kernel of f is 0, and f is an isomorphism,

THEOREM 13. Let I be a non-zero ideal of R. Then

(1) $HI \cong H \otimes_R I$ is the completion of I in the R-topology.

(2) $H/HI \cong R/I$ and $HI/I \cong H/R$.

(3) $R \cap HI = I$ and $R + HI = H$.

Proof. We define an epimorphism $f: H \otimes_R I \to HI$ by $f(h \otimes x) = hx$ for $h \in H$ and $x \in I$; and we let $i: H \otimes_R R \to H$ be the canonical isomorphism. Then we have a commutative diagram

$$0 \to H \otimes_R I \to H \otimes_R R \to H \otimes_R R/I \to 0$$

with vertical maps f and i

$$0 \to HI \to H \to H/HI \to 0$$

By Theorem 10, H is a flat R-module; and thus the rows of the diagram are exact. Since i is an isomorphism, it follows that f is an isomorphism and that there is an induced isomorphism

$g: H \otimes_R R/I \to H/HI$. By Theorem 11, $H \otimes_R R/I \cong R/I$. Thus we have $H/HI \cong R/I$.

We have an exact sequence:

$$0 \to R/I \to H/I \to H/R \to 0$$

Since H/R is torsion-free and divisible by Theorem 10, and R/I is a cotorsion module, this sequence splits. Therefore, $H/I = R/I \oplus B/I$, where B is a submodule of H containing I and $B/I \cong H/R$. We have $R + B = H$ and $R \cap B = I$. We will show that $B = HI$ and this will prove statements (2) and (3).

Now we have an exact sequence:

$$0 \to HI/I \to H/I \xrightarrow{\phi} H/HI \to 0 .$$

Since $H/HI \cong R/I$, and B/I is torsion-free and divisible, we have $B/I \subset \mathrm{Ker}\, \phi = HI/I$ and $\phi(R/I) = H/HI \cong R/I$. By Theorem 12, it follows that $R/I \cap \mathrm{Ker}\, \phi = 0$. Since $H/I = R/I \oplus B/I$, we see that $B/I = HI/I$. Therefore, $B = HI$; and statements (2) and (3) are proved.

Since H/HI is isomorphic to R/I, it is a cotorsion R-module; furthermore, H is also a cotorsion R-module. Therefore, it follows from Theorem 1 that HI is a cotorsion R-module. Since HI/I is torsion-free and divisible, we have $K \otimes_R I \cong K \otimes_R HI$. Thus by Theorem 9 and Theorem 6 we have $\tilde{I} \cong \operatorname{Hom}_R(K, K \otimes_R I) \cong \operatorname{Hom}_R(K, K \otimes_R HI) \cong HI$. Thus HI is the completion of I in the R-topology.

THEOREM 14. Let I be a non-zero ideal of R. Then I is a cotorsion R-module if and only if R is a complete ring.

Proof. Since R/I is a cotorsion R-module, it follows from Theorem 1 that I is a cotorsion R-module if and only if R is a cotorsion R-module. But R is cotorsion if and only if R is complete in the R-topology by Theorem 9.

The following important theorem shows that we can change rings without changing the topology on a module.

THEOREM 15. If S is a ring extension of R in Q, and if A is an S-module, then the S-topology and the R-topology on A are the same. Consequently, if A is a torsion-free S-module, then A is complete in the S-topology if and only if A is complete in the R-topology; that is, A is S-cotorsion if and only if A is R-cotorsion.

Proof. If $r \epsilon R$, $r \neq 0$, then $r \epsilon S$; and hence rA is an open set in the S-topology on A. On the other hand, if $s \epsilon S$, $s \neq 0$, then $s = a/b$, where $a, b \epsilon R$ and $b \neq 0$. Thus $aA = sbA \subset sA$, and this implies that sA is an open set in the R-topology on A. Therefore, the two topologies are the same. The last statement of the theorem now follows from Theorem 9.

Theorem 16 gives an important criterion for an homomorphic image of Q to be indecomposable.

THEOREM 16. If A is a non-zero R-submodule of Q, then A is complete in the R-topology if and only if $\operatorname{Hom}_R(Q, Q/A) \cong Q$. Thus if A is complete, then Q/A is an indecomposable R-module.

Proof. Without loss of generality we can assume that A is reduced; that is, $A \neq Q$. Since $Q \cong \operatorname{Hom}_R(Q, Q)$ we have an exact sequence

$$0 \to Q \to \operatorname{Hom}_R(Q, Q/A) \to \operatorname{Ext}^1_R(Q, A) \to 0.$$

By Theorem 9, A is complete in the R-topology if and only if $\operatorname{Ext}^1_R(Q, A) = 0$; and this proves the first statement. The second statement follows from the first and exact sequence (III).

h-LOCAL RINGS

DEFINITION.   R is said to be a <u>quasi-local ring</u> if R has only one maximal ideal; and R is said to be a <u>quasi-semilocal ring</u> if R has only a finite number of maximal ideals.

THEOREM 17.   Let R be a quasi-local domain with maximal ideal M. Then H is a commutative quasi-local ring with maximal ideal HM, and $H/HM \cong R/M$.

<u>Proof</u>.   By Theorem 10, H is a commutative ring; and by Theorem 13, $H/HM \cong R/M$. Therefore, HM is a maximal ideal of H. Suppose that J is an ideal of H that is not contained in HM. Then $H = HM + J$, and so there exist $g \in HM$ and $f \in J$ such that $1 = g + f$. Now we have $g = \sum_{i=1}^{n} h_i m_i$, where $h_i \in H$ and $m_i \in M$ for $i = 1, \ldots, n$.

Let x be an element of K; then as in the proof of Theorem 10, we have $h_i(Rx) \subset Rx$ for all i. Thus there exist elements $r_i \in R$ such that $h_i(x) = r_i x$ for all i. Therefore, $g(x) = \sum_{i=1}^{n} h_i(m_i x) = \sum_{i=1}^{n} r_i m_i x = mx$, where $m \in M$. Hence

$$f(x) = x - g(x) = x - mx = (1-m)x .$$

Let $u = 1 - m$; then u is a unit in R and we have $f(x) = ux$. It follows that f is an automorphism of K, and thus $f^{-1} \in H$. Therefore, $J = H$; and we see that H is a quasi-local ring with maximal ideal HM.

THEOREM 18. Let R be a quasi-local ring. Then K is an indecomposable R-module.

Proof. Since H is a quasi-local ring by Theorem 17, it can have no idempotents other than the identity. Therefore, K is an indecomposable R-module.

DEFINITION. R is said to be an h-local ring, if it satisfies the following two conditions:

(1) Each non-zero prime ideal of R is contained in only one maximal ideal of R.

(2) Each non-zero element of R is contained in only a finite number of maximal ideals of R.

Thus R is h-local if and only if modulo any non-zero prime ideal it is a quasi-local ring, and modulo any non-zero ideal it is a quasi-semilocal ring.

The notion of an h-local ring has proved to be extremely useful, and it crops up under an amazing variety of conditions. We will explore a number of equivalent definitions of this kind of ring. It is this richness of description which gives the notion its tremendous power. But first we need some lemmas.

THEOREM 19. Let M and N be maximal ideals of R. Then the following statements are equivalent:

(1) $R_M \otimes_R R_N \cong Q$.

(2) $M \cap N$ does not contain a non-zero prime ideal of R.

Proof. Let S be the multiplicatively closed set R - N. Then $R_M \otimes_R R_N = (R_M)_S$. We have that $(R_M)_S = Q$ if and only if $(R_M)_S$ is a field, if and only if every nonzero prime ideal of R_M meets S, if and only if $M \cap N$ does not contain a nonzero prime ideal of R.

NOTATION. Let M be a maximal ideal of R, and A an R-module. Then we will denote $R_M \otimes_R A$ by A_M and $\text{Hom}_R(R_M, A)$ by A^M.

THEOREM 20. Let M and N be maximal ideals of R and suppose that $R_M \otimes_R R_N \cong Q$. Then

(1) If A is a torsion R-module, and B an R-module, we have $\text{Hom}_R(A_M, B_N) = 0$.

(2) If C is an h-reduced R-module, and B an R-module, we have $\text{Hom}_R(B^M, C^N) = 0$.

Proof. (1) Since $\text{Hom}_R(R_M, B_N)$ is both an R_M-module and an R_N-module, it is an $R_M \otimes_R R_N$-module. But $R_M \otimes_R R_N \cong Q$, and thus $\text{Hom}_R(A, \text{Hom}_R(R_M, B_N)) = 0$, since A is a torsion module. By Theorem A1 we have

$$\text{Hom}_R(A_M, B_N) \cong \text{Hom}_R(A, \text{Hom}_R(R_M, B_N)) .$$

(2) Now $B^M \otimes_R R_N$ is both an R_M-module and an R_N-module, and hence it is an $R_M \otimes_R R_N$-module. But $R_M \otimes_R R_N \cong Q$, and since C is an h-reduced module, we have $\text{Hom}_R(B^M \otimes_R R_N, C) = 0$. By Theorem A1 we have

$$\text{Hom}_R(B^M, C^N) \cong \text{Hom}_R(B^M \otimes_R R_N, C) .$$

THEOREM 21. Suppose that $K = A \oplus B$ is the direct sum of two submodules A and B. Then $\text{Hom}_R(A, B) = 0$.

Proof. Let $f \in \text{Hom}_R(A, B)$. Then f can be extended to a homomorphism of K into K by sending B into 0. Let $x \in A$; then as in the proof of Theorem 10 we have $f(Rx) \subset Rx$. Therefore $f(A) \subset A$. But by definition $f(A) \subset B$. Since $A \cap B = 0$, we have that $f = 0$.

NOTATION. If M is a maximal ideal of R, we will denote by $[M]$ the intersection of all the rings R_N, where N is a maximal ideal different from M. If R is a quasi-local ring with maximal ideal M, then we will let $[M] = Q$.

Let A be an R-module; then $\sum_M \oplus A_M$ will denote the direct sum of the modules A_M, where M ranges over all of the maximal ideals of R. Similarly $\prod_M A^M$ will denote the direct product of the modules A^M. For each M we will let $H(M) = \text{Hom}_R(K_M, K_M)$, the completion of R_M in the R- (or R_M)-topology.

We note for further use that if A and B are R_M-modules, then $\text{Hom}_R(A, B) \cong \text{Hom}_{R_M}(A, B)$.

THEOREM 22. The following statments are equivalent:

(1) R is an h-local ring.

(2) $[M] \otimes_R R_M \cong Q$ for every maximal ideal M of R.

(3) $K \cong \sum_M \oplus K_M$.

(4) $T \cong \sum_M \oplus T_M$ for every torsion R-module T.

(5) $H \cong \prod_M H(M)$.

(6) $C \cong \prod_M C^M$ for every cotorison R-module C.

Proof. (1) $\Rightarrow$ (2). Let M be a maximal ideal of R. Let $A = \prod_{N \neq M} K_N$ and $B = \sum_{N \neq M} \oplus K_N$, where N ranges over the maximal ideals of R different from M. Define $\phi: Q \to A$ by $\phi(x) = \langle x + R_N \rangle$

for $x \in Q$. Now $x = a/b$, where $a, b \in R$ and $b \neq 0$. Since b is contained in only a finite number of maximal ideals of R, we have $x \in R_N$ for all but a finite number of maximal ideals N. Hence we actually have $\text{Im } \emptyset \subset B$. Since $\text{Ker } \emptyset = [M]$, we have an exact sequence:
$$0 \to Q/[M] \to B.$$

Now if N is a maximal ideal of R different from M, we have by Theorem 19 that $R_M \otimes_R R_N \cong Q$. Thus $R_M \otimes_R K_N = 0$, from which it follows that $B_M = 0$. Since $Q/[M] \subset B$, we have
$$[M] \otimes_R R_M \cong Q.$$

$(2) \Rightarrow (3)$. Let M be a maximal ideal of R, and let $A = \sum_{N \neq M} [N]$, the sum (not direct) of all of the $[N]$ for N a maximal ideal different from M. Then $R \subset A \subset R_M$, and hence $A_M = R_M$. On the other hand, by assumption, we have $A_N = Q$ for all $N \neq M$. Thus $A = \bigcap_{N \neq M} A_N \cap A_M = Q \cap R_M = R_M$. Consequently, we have $(R_M + [M])_N = (A + [M])_N = Q$ for all maximal ideals N of R including M. Therefore, $R_M + [M] = Q$. Since $R_M \cap [M] = R$, we have
$$K_M = Q/R_M = (R_M + [M])/R_M \cong [M]/(R_M \cap [M]) = [M]/R.$$

Now we have $Q = A + [M] = \sum_N [N]$, where the sum ranges over all maximal ideals N of R including M. If $N_1, \ldots, N_t$ is any finite set of maximal ideals of R different from M, then $R \subset (\sum_{i=1}^{t} [N_i] \cap [M]) \subset A \cap [M] = R$. Thus $(\sum_{i=1}^{t} [N_i] \cap [M]) = R$. From these facts it follows that $K = Q/R = \sum_N \oplus ([N]/R)$, where N ranges over all maximal ideals of R. Since we have shown that $[N]/R \cong K_N$, we have $K \cong \sum_N \oplus K_N$.

$(3) \Rightarrow (4)$. If T is a torsion R-module, we have $\text{Tor}_1^R(K_M, T) \cong T_M$, for every maximal ideal of R. Thus $T \cong \text{Tor}_1^R(K, T) \cong \text{Tor}_1^R(\sum_M \oplus K_M, T) \cong \sum_M \oplus \text{Tor}_1^R(K_M, T) \cong \sum_M \oplus T_M$.

$(4) \Rightarrow (5)$. Since $K \cong \sum_M \oplus K_M$, we have by Theorem 21 that $\text{Hom}_R(K_M, K_N) = 0$ for every pair of distinct maximal ideals M, N of R. Thus $H = \text{Hom}_R(K, K) \cong \text{Hom}_R(\sum_M \oplus K_M, K) \cong \prod_M \text{Hom}_R(K_M, K)$

$\cong \prod_M \text{Hom}_R(K_M, K_M) \cong \prod_M H(M)$.

$(5) \Rightarrow (6)$. Let $\Sigma = \sum_M \oplus H(M)$; since $H = \prod_M H(M)$, we have $\Sigma \subset H$, and we let $U = H/\Sigma$. Clearly U is a torsion-free R-module. We will prove that U is a divisible R-module. Let M be a maximal ideal of R and let $A = \sum_{N \neq M} \oplus H(N)$ and $B = \prod_{N \neq M} H(N)$, where N ranges over all maximal ideals of R different from M. Then $\Sigma = A \oplus H(M)$ and $H = B \oplus H(M)$, and thus $B/A \cong U$.

Let I be a nonzero ideal of R. Since $H_M \cong H(M) \oplus B_M$, we have by Theorem 11 that $R_M/I_M \cong H \otimes_R R_M/I_M \cong H_M \otimes_R R_M/I_M$ $\cong (H(M) \oplus B_M) \otimes_R R_M/I_M \cong R_M/I_M \oplus B_M/IB_M$. Thus we have an epimorphism of R_M/I_M onto R_M/I_M with kernel isomorphic to B_M/IB_M. Since such an epimorphism is an isomorphism by Theorem 12, we have $B_M = IB_M$. Therefore, B_M is a divisible R-module. Now $U_M \cong B_M/A_M$, and hence U_M is torsion-free and divisible for every maximal ideal M of R. Thus U is a torsion-free and divisible R-module.

Let C be a cotorsion R-module. Since $U = H/\Sigma$ is torsion-free and divisible, we have $\text{Hom}_R(H, C) \cong \text{Hom}_R(\Sigma, C)$. But $\text{Hom}_R(H, C) \cong C$ by Theorem 11; and $\text{Hom}_R(\Sigma, C) \cong \prod_M \text{Hom}_R(H(M), C)$ $\cong \prod_M \text{Hom}_R(R_M, C) = \prod_M C^M$. Therefore $C \cong \prod_M C^M$.

(6) $\Rightarrow$ (1). Let P be a nonzero prime ideal of R. If R/P is a decomposable R-module, there exist two ideals I and J of R, properly containing P, with $I \cap J = P$; and this is a contradiction. Thus R/P is an indecomposable cotorsion R-module. Hence, by assumption, there exists a maximal ideal M such that $R/P \cong (R/P)^M$. Therefore R/P is an R_M-module which implies that $R/P \cong (R/P)_M$ $\cong R_M/P_M$. Since $P_M \neq R_M$, we have $P \subset M$.

Let N be a maximal ideal of R different from M. Then there exists an element $s \in N$ such that $s \in R - M$. This element s acts as a unit on R_M/P_M which is isomorphic to R/P; and hence there exists $r \in R$ such that $s(r + P) = 1 + P$. Since we then have $1 - sr \in P$, it follows that P is not contained in N. Therefore P is contained in only one maximal ideal of R.

Let I be a nonzero ideal of R, and let $T = R/I$. Since T is a cotorsion R-module, we have, by assumption, that $T \cong \prod_M T^M$. Let $S = \sum_M \oplus T^M \subset T$; since S is a cotorsion R-module, we have $S \cong \prod_M S^M$. Let M and N be distinct maximal ideals of R. Then by Theorems 19 and 20 we have $\text{Hom}_R(R_M, T^N) = 0$. Therefore, $S^M = \text{Hom}_R(R_M, \sum_N \oplus T^N) \cong \text{Hom}_R(R_M, T^M) \cong T^M$. Thus $S \cong \prod_M S^M \cong \prod_M T^M \cong T$, and hence $S \cong R/I$.

Now for $N \neq M$ we have by Theorem 19 that

$$(T^N)_M \cong T^N \otimes_R Q = 0. \quad \text{Thus} \quad S_M \cong \sum_N \oplus (T^N)_M = (T^M)_M = T^M.$$

Therefore $S = \sum_M \oplus T^M \cong \sum_M \oplus S_M$. Since $S \cong R/I$, we have

$S_M \cong R_M/I_M$. Hence $R/I \cong \sum_M \oplus R_M/I_M$. Since R/I is a cyclic

R-module, it follows that $I_M = R_M$ for all but a finite number of

maximal ideals M. Therefore I is contained in only a finite number

of maximal ideals.

We have proved that R is an h-local ring.

THEOREM 23. If R is an h-local ring, then H is a direct

product of commutative, quasi-local rings.

Proof. This is an immediate consequence of Theorems 17

and 22.

The following theorem presents one of the most useful aspects

of h-local rings and we will make frequent use of it.

THEOREM 24. If R is an h-local domain, and C an

R-module, then inj. dim$_R C = \sup_M$ inj. dim$_{R_M} C_M$, where M ranges

over all maximal ideals of R. Thus gl. dim R $= \sup_M$ gl. dim R$_M$.

Proof. We will prove first that C is an injective R-module

if and only if C_M is an injective R_M-module for every maximal ideal

M of R. Suppose that C is an injective R-module. If T is the

torsion submodule of C, then $C = T \oplus S$, where S is torsion-free

and divisible. Thus $C_M = T_M \oplus S$, and it is sufficient to prove that

T_M is an injective R_M-module. But T_M is a direct summand of T by Theorem 22 and hence T_M is an injective R-module. It now follows readily from Theorem A3 that T_M is an injective R_M-module.

Conversely, suppose that C_M is an injective R_M-module for every M. Again let T be the torsion submodule of C, and this time let $J = C/T$. Then T_M is the torsion submodule of C_M, and we have $C_M \cong T_M \oplus S_M$. Therefore, S_M is torsion-free and divisible for every M, and thus S is torsion-free and divisible. It is thus sufficient to prove that T is an injective R-module.

Let I be a nonzero ideal of R and take an exact sequence of the form:

$$(1) \qquad 0 \to T \to B \to R/I \to 0 .$$

Then B is a torsion R-module. Since R is an h-local ring, we have $T = \sum_M \oplus T_M$ and $B = \sum_M \oplus B_M$ by Theorem 22. By Theorem 20, $T_M \subset B_M$ for every M. Since T_M is a direct summand of C_M, it is an injective R_M-module. Thus T_M is a direct summand of B_M. It follows that T is a direct summand of B, and thus exact sequence (1) splits. Therefore $\mathrm{Ext}_R^1(R/I, T) = 0$. This proves that T is an injective R-module, and hence $C \cong T \oplus S$ is also an injective R-module.

Now assume that C is an arbitrary R-module and consider an exact sequence of the form:

$$(2) \qquad 0 \to C \to E_0 \to E_1 \to \cdots \to E_{n-1} \to D \to 0,$$

where E_i is an injective R-module for all i, $0 \leq i \leq n-1$, and D is an R-module. Since R_M is a flat R-module, we obtain an exact sequence:

$$(3) \qquad 0 \to C_M \to E_{0_M} \to E_{1_M} \to \cdots \to E_{n-1_M} \to D_M \to 0 .$$

By what we have already shown, E_{i_M} is an injective R_M-module for every M. Furthermore D is an injective R-module if and only if D_M is an injective R_M-module for every M. It follows readily that $\operatorname{inj\,dim}_R C = \sup \operatorname{inj.\,dim}_{R_M} C_M$.

THEOREM 25. Let I be an ideal of R that is contained in only one maximal ideal M of R. Then $I_M \cap R = I$ and $R/I \cong R_M/I_M$.

Proof. Clearly $I \subset I_M \cap R$. On the other hand let $x \in I_M \cap R$; then $x = a/s$, where $a \in I$ and $s \in R - M$. Since $Rs + I = R$, there exist elements $r \in R$ and $b \in I$ such that $1 = rs + b$. Thus $x = rsx + bx = ra + bx \in I$. Therefore, $I = I_M \cap R$.

Let N be any maximal ideal of R different from M. Since I is not contained in N, we have $I_N = R_N$. Thus $(I_M + R)_N \supset (I_M)_N = (R_N)_M \supset R_M$. We also have $(I_M + R)_M = R_M$. Thus $I_M + R = \bigcap_{N \neq M} (I_M + R)_N \cap (I_M + R)_M = R_M$. Therefore,

$$R_M/I_M = (I_M + R)/I_M \cong R/(R \cap I_M) = R/I .$$

The following theorem is essentially due to I. S. Cohen [6].

THEOREM 26. Let R be a ring such that every nonzero ideal is contained in only a finite number of maximal ideals of R. Then

(1) R is Noetherian if and only if R_M is Noetherian for every maximal ideal M of R.

(2) Let I be a nonzero ideal of R, and let k be the supremum over all maximal ideals M of R of the minimum number of

generators required to generate I_M as an R_M-ideal. Then I can be generated as an ideal of R by $\leq \max(2, k)$ generators.

Proof. If R is a Noetherian ring, then of course R_M is Noetherian for every maximal ideal M of R. Now $I_M = R_M$ for all but a finite number of maximal ideals M. Thus, if we prove the second statement of the theorem, we will also have proved the first. We can assume without loss of generality that $k < \infty$, and we let $n = \max(2, k)$.

Let $M_1, \ldots, M_t$ be the maximal ideals of R that contain I. For each i, $1 \leq i \leq t$, let a_{1i} be an element of I that is an element of a minimal generating set for I_{M_i}. By the Chinese Remainder Theorem there is an element $a_1 \in I$ such that $a_1 \equiv a_{1i} \pmod{M_i I}$ for all i. Let $N_1, \ldots, N_q$ be the maximal ideals of R that contain a_1. Then for each j, $1 \leq j \leq q$, we may choose $a_{2j}, \ldots, a_{nj} \in I$ so that $(a_1, a_{2j}, \ldots, a_{nj})$ generate I_{N_j} over R_{N_j}. For if N_j is one of the M_i's, then a_1 is part of a minimal generating set for I_{M_i}; while if N_j is not one of the M_i's, then $I_{N_j} = R_{N_j}$ and hence I_{N_j} can be generated by a single element over R_{N_j}.

Using the Chinese Remainder Theorem again we may choose for each p, $2 \leq p \leq n$, an element $a_p \in I$ such that $a_p \equiv a_{pj} \pmod{N_j I}$ for all $j = 1, \ldots, q$. If we let J be the ideal of R generated by $a_1, a_2, \ldots, a_n$, then $J_{N_j} = I_{N_j}$ for $j = 1, \ldots, q$. If N is a maximal ideal of R different from any of the N_j's, then $a_1 \notin N$, and thus $J_N = R_N = I_N$. Therefore, J and I agree locally and hence are equal. This concludes the proof of the theorem.

4

REFLEXIVE RINGS

DEFINITIONS. Let A be an R-module and denote $\mathrm{Hom}_R(A,R)$ by A'. Then we have a canonical map $\lambda: A \to A''$ defined by

$$[\lambda(x)](f) = f(x)$$

for all $x \in A$ and $f \in A'$. A is called a <u>torsionless</u> R-module if λ is a monomorphism, and a <u>reflexive</u> R-module if λ is an isomorphism.

A torsionless R-module is of course torsion-free. It is not hard to see that A is torsionless if and only if A is isomorphic to a submodule of a direct product of copies of R. Finitely generated projective R-modules provide examples of reflexive modules, although there are many others.

Let B be a nonzero R-submodule of Q and define $B^{-1} = \{q \in Q \mid qB \subseteq R\}$. Every R-homomorphism from B to R can be extended to an R-homomorphism from Q to Q. If we let $q = f(1)$, then f is multiplication on Q by q. Thus we have a canonical isomorphism of B^{-1} and $\mathrm{Hom}_R(B, R)$.

B is called a <u>fractionary ideal</u> of R if $B^{-1} \neq 0$. It is easy to see that B is a fractionary ideal of R if and only if B is isomorphic to an ideal of R. A fractionary ideal is torsionless; and if we identify B^{-1} with $\mathrm{Hom}_R(B, R)$ in this case, then the inclusion map $B \subseteq B^{-1-1}$ becomes identified with the mapping λ defined earlier.

We define the <u>rank</u> of a torsion-free R-module A to be the dimension over Q of the Q-vector space $Q \otimes_R A$. The rank of A is equal to the cardinality of a basis of a maximal free R-submodule contained in A.

If A is a torsion-free R-module of finite rank, then A' is torsion-free and rank A' ≤ rank A. For let F be a maximal free submodule of A. Then rank F = rank A and A/F is a torsion R-module. Thus we have an exact sequence $0 \to A' \to F'$. Since rank F = rank F', we have rank A' ≤ rank A. The question of when rank A' = rank A is answered in the next theorem.

THEOREM 27. Let A be a torsion-free R-module of finite rank. Then the following statements are equivalent.

    (1) A is torsionless.

    (2) A is a submodule of a finitely generated torsion-free R-module.

    (3) rank A = rank A".

    (4) rank A = rank A'.

<u>Proof.</u> (1) $\Rightarrow$ (2). Let F be a maximal free R-submodule of A'. Then F is finitely generated and A'/F is a torsion R-module. Thus we have a monomorphism: $A'' \to F'$. Since F' is isomorphic to F, and since $\lambda : A \to A''$ is a monomorphism, A is isomorphic to a submodule of a finitely generated free R-module.

(2) $\Rightarrow$ (1) Because finitely generated torsion-free modules can be embedded in finitely generated free modules, A is a submodule of a free R-module. Thus if x is a non-zero element of A, there is a

homomorphism $f: A \to R$ such that $f(x) \neq 0$. This shows that the map λ is a monomorphism.

$(1) \Rightarrow (3)$. Since λ is a monomorphism, we have rank $A \leq$ rank A''. However, we always have rank $A'' \leq$ rank A' $\leq$ rank A. Thus rank $A =$ rank A''.

$(3) \Rightarrow (4)$. We have rank $A'' \leq$ rank $A' \leq$ rank A. If rank $A'' =$ rank A, then rank $A' =$ rank A also.

$(4) \Rightarrow (1)$. Let F be a maximal free R-submodule of A. Then rank $F =$ rank A, and A/F is a torsion module. Thus we have a monomorphism $A' \to F'$. Since rank $A =$ rank A' and rank $F =$ rank F', it follows that F'/A' is a torsion module. Thus we have a commutative diagram

$$
\begin{array}{ccccc}
0 & \to & F & \to & A \\
 & & \lambda_F \downarrow & & \lambda_A \downarrow \\
0 & \to & F'' & \to & A'' \; .
\end{array}
$$

Let $B = \operatorname{Ker} \lambda_A$; then $B \cap F$ goes to 0 under the composite $F \xrightarrow{\lambda_F} F'' \longrightarrow A''$. But this composite is a monomorphism, and thus $B \cap F = 0$. Since A/F is a torsion R-module, we have $B = 0$. Thus A is a torsionless R-module.

The following corollary of Theorem 27 shows that there is an abundancy of reflexive R-modules.

THEOREM 28. If A is a torsion-free R-module of finite rank, then A' is a reflexive R-module.

Proof. Let $\lambda_A: A \to A''$ and $\lambda_{A'}: A' \to A'''$ be the canonical R-homomorphisms. Then λ_A induces a homomorphism $\mu: A''' \to A'$.

It is easily verified that the composite $\mu\lambda_{A'}$ is the identity on A'. Thus A' is isomorphic to a direct summand of A''' and $\lambda_{A'}$ is a monomorphism. But then A' is a torsionless R-module, and hence by Theorem 27 we have rank $A' = $ rank A'''. Therefore A' can not be a proper direct summand of A''', and thus $\lambda_{A'}$ is an isomorphism.

DEFINITIONS. An R-module will be called a <u>universal injective</u> R-module if it is injective and contains a copy of every simple R-module.

We will let $\mathcal{S}$ denote the direct sum of one copy of every simple R-module, and $E(\mathcal{S})$ the injective envelope of $\mathcal{S}$. Then $E(\mathcal{S})$ is a universal injective R-module.

An integral domain R is called a <u>reflexive ring,</u> if every torsionless R-module of finite rank is reflexive. By Theorem 27 this is equivalent to saying that every submodule of a finitely generated, torsion-free R-module is reflexive. The connection with universal injectives is provided by the next theorem.

THEOREM 29. The following statements are equivalent:

(1) R is a reflexive ring.

(2) K is a universal injective R-module.

(3) $K \cong E(\mathcal{S})$.

(4) Every ideal of R is reflexive and K is injective.

<u>Proof.</u> (1) $\Rightarrow$ (2) Let I be a nonzero ideal of R. We will show that $\operatorname{Ext}_R^1(I, R) = 0$. Consider an extension of R by I:

(a) $$0 \longrightarrow R \xrightarrow{\alpha} A \xrightarrow{\beta} I \longrightarrow 0 .$$

Clearly A is a torsion-free R-module of rank 2. We will show first that A is a torsionless R-module. We have a derived exact sequence obtained by applying the functor $\mathrm{Hom}_R(\cdot, R)$ to exact sequence (a):

(b) $\qquad 0 \longrightarrow I' \xrightarrow{\beta'} A' \xrightarrow{\alpha'} R' \longrightarrow \mathrm{Ext}_R^1(I, R)$.

Now $\mathrm{Ext}_R^1(I, R)$ is a torsion R-module, since it is isomorphic to $\mathrm{Ext}_R^2(R/I, R)$. Hence it follows from (b) that rank A' = rank I' + rank R' = 2. But then rank A = rank A', and so A is torsionless by Theorem 27.

Let $J = \mathrm{Im}\ \alpha'$ in exact sequence (b). We have just shown that J is not zero, and thus it is isomorphic to a non-zero ideal of R. We now have a commutative diagram with exact rows:

(c)
$$
\begin{array}{ccccccccc}
0 & \longrightarrow & R & \xrightarrow{\alpha} & A & \xrightarrow{\beta} & I & \longrightarrow & 0 \\
& & & & \lambda_A \downarrow & & \lambda_I \downarrow & & \\
0 & \longrightarrow & J' & \longrightarrow & A'' & \xrightarrow{\beta''} & I'' & &
\end{array}
$$

Since A and I are torsionless of finite rank, we have by assumption that λ_A and λ_I are in fact isomorphisms. It follows that β'' is onto and that λ_A induces an isomorphism $\gamma: R \to J'$. Thus J is isomorphic to R. But this means that the exact sequence

(d) $\qquad 0 \to I' \to A' \xrightarrow{\alpha'} J \to 0$

splits. Hence the bottom row of (c) is exact and splits. Therefore the top row of (c) splits. This shows that $\mathrm{Ext}_R^1(I, R) = 0$.

Since $\mathrm{Ext}_R^2(R/I, R) \cong \mathrm{Ext}_R^1(I, R) = 0$ for every ideal I of R, we have inj.$\dim_R R = 1$; that is, K is injective. It remains to be proved that K is universal. Let M be a maximal ideal of R. Since R is a reflexive ring, $M^{-1} \neq R$; and thus M^{-1}/R is a non-zero sub-module of K. But M^{-1}/R is annihilated by M, and hence is a direct sum of copies of R/M. Thus K contains a copy of every simple R-module, and hence K is a universal injective R-module.

(2) $\Rightarrow$ (3). Since K is a universal injective R-module, K contains a copy of $E(\mathcal{J})$ as a direct summand. Thus there exists a submodule B of K such that $K \cong E(\mathcal{J}) \oplus B$. By Theorem 21 we have $\operatorname{Hom}_R(B, E(\mathcal{J})) = 0$. But $E(\mathcal{J})$ is a universal injective R-module, and hence we must have $B = 0$.

(3) $\Rightarrow$ (4). Let I be a non-zero ideal of R. We only need to prove that I is reflexive. Suppose that $I^{-1-1} \neq I$, and choose an element $a \in I^{-1-1}$ such that $a \notin I$. Then $x = a + I$ is a non-zero element of R/I. Since K is a universal injective, there exists an R-homomorphism $f: R/I \to K$ such that $f(x) \neq 0$. Since I annihilates Im f, we have Im $f \subset I^{-1}/R$. Thus I^{-1-1} annihilates Im f, and we have $f(x) = af(1 + I) = 0$. This contradiction shows that I is a reflexive ideal.

(4) $\Rightarrow$ (1). Let A be a torsionless R-module of finite rank $n < \infty$. Then there exists a homomorphism of A onto a non-zero ideal I of R with kernel B of rank n-1. Since inj. $\dim_R R = 1$, we have $\operatorname{Ext}_R^1(I, R) = 0$. Thus we have an exact sequence:

$$0 \to I' \to A' \to B' \to 0 .$$

Therefore, rank B' = rank A' - rank I' = n-1 = rank B. Hence B and B' are torsionless R-modules of rank n-1 by Theorem 27. Since B' is a submodule of a free R-module by the same theorem, we have $\operatorname{Ext}_R^1(B', R) = 0$. Therefore, we have a commuative diagram with exact rows:

$$0 \to B \to A \to I \to 0$$
$$\lambda_B \downarrow \quad \lambda_A \downarrow \quad \lambda_I \downarrow$$
$$0 \to B'' \to A'' \to I'' \to 0$$

Now λ_I is an isomorphism, since I is reflexive; and λ_B is an isomorphism by induction on n = rank A. Therefore λ_A is an isomorphism, and A is a reflexive R-module. Thus R is a reflexive ring.

THEOREM 30. Let R be a reflexive ring. Then R is an h-local ring.

Proof. Let $\{M_\gamma\}$, $\gamma \in \Gamma$, be the collection of all of the maximal ideals of R. Let $\mathcal{J}_\gamma = R/M_\gamma$, and let $\mathcal{J} = \sum_{\gamma \in \Gamma} \oplus \mathcal{J}_\gamma$. Then by Theorem 29 we can assume that $\mathcal{J} \subset K$ and that $K = E(\mathcal{J})$. Let E_γ be the injective envelope of $\mathcal{J}_\gamma$ in K. If $x \in K$, we define $O(x)$ to be the R-annihilator of x; that is $O(x) = \{r \in R \mid rx = 0\}$.

We will prove a number of statements which will culminate in the proof of the theorem.

(1) If $x \in E_\gamma$, $x \neq 0$, then $O(x) \subset M_\gamma$

Since E_γ is an essential extension of $\mathcal{J}_\gamma$, there exists an element $r \in R$ such that $rx \neq 0$ and $rx \in \mathcal{J}_\gamma$. Thus $M_\gamma rx = 0$. If $O(x)$ is not contained in M_γ, choose an element $s \in R - M_\gamma$ such that $sx = 0$. But then $Rrx = (M_\gamma + Rs)rx = 0$. This contradiction shows that $O(x) \subset M_\gamma$.

(2) <u>If</u> $s \in R - M_\gamma$, <u>then multiplication by</u> s <u>is an automorphism of</u> E_γ.

It follows from (1) that multiplication by s is a monomorphism of E_γ. Since E_γ is divisible, it follows that multiplication by s is an epimorphism of E_γ.

(3) <u>There exists an R-submodule</u> D_γ <u>of</u> K <u>such that</u> $K = E_\gamma \oplus D_\gamma$.
<u>We have</u> $E_\gamma \cong K_{M_\gamma}$, <u>and</u> $(D_\gamma)_{M_\gamma} = 0$.

Since E_γ is an injective R-module, there exists an R-submodule D_γ of K such that $K = E_\gamma \oplus D_\gamma$. It follows from (2) that $(E_\gamma)_{M_\gamma} \cong E_\gamma$. Thus we have $K_{M_\gamma} \cong E_\gamma \oplus (D_\gamma)_{M_\gamma}$. Since R_{M_γ} is a quasi-local ring, we have by Theorem 18 that K_{M_γ} is an indecomposable R_{M_γ}-module. Thus $(D_\gamma)_{M_\gamma} = 0$ and $K_{M_\gamma} \cong E_\gamma$.

(4) $D_\gamma = \{x \in K \mid sx = 0 \ \underline{\text{for some}} \ s \in R - M_\gamma\}$.

Let $A_\gamma = \{x \in K \mid sx = 0 \text{ for some } s \in R - M_\gamma\}$. It is clear that A_γ is an R-submodule of K. Since $(D_\gamma)_{M_\gamma} = 0$ by (3), we have $D_\gamma \subseteq A_\gamma$. By (1) we have $E_\gamma \cap A_\gamma = 0$. Let $x \in A_\gamma$; then $x = x_\gamma + y_\gamma$, where $x_\gamma \in E_\gamma$ and $y_\gamma \in D_\gamma \subseteq A_\gamma$. Thus $x_\gamma \in E_\gamma \cap A_\gamma = 0$, and so $x = y_\gamma \in D_\gamma$. Therefore, we have $D_\gamma = A_\gamma$.

(5) <u>Let</u> M_δ <u>be a maximal ideal different from</u> M_γ. <u>Then</u> $E_\delta \subseteq D_\gamma$ <u>and</u> $E_\gamma \cap E_\delta = 0$.

Let $x \in E_\delta$; then $x = x_\gamma + y_\gamma$, where $x_\gamma \in E_\gamma$ and $y_\gamma \in D_\gamma$. Suppose that $x_\gamma \neq 0$. By (4) there exists $s \in R - M_\gamma$ such that $sy_\gamma = 0$. Hence $sx = sx_\gamma \in E_\gamma \cap E_\delta$, and by (2) $sx_\gamma \neq 0$. Since E_δ is an essential extension of $\mathcal{A}_\delta$, there exists an element $r \in R$ such that $rsx_\gamma \in \mathcal{A}_\delta$ and $rsx_\gamma \neq 0$. Therefore, $M_\delta rsx_\gamma = 0$; and thus by (1) $M_\delta \subseteq O(rsx_\gamma) \subseteq M_\gamma$. But then $M_\delta = M_\gamma$, and this contradiction shows that $x_\gamma = 0$. Thus $x = y_\gamma \in D_\gamma$, and so $E_\delta \subset D_\gamma$. Therefore by (3) $E_\gamma \cap E_\delta = 0$.

(6) $E_\gamma \cap (E_{\delta_1} + \ldots + E_{\delta_n}) = 0$ if $M_{\delta_k} \neq M_\gamma$ for $k = 1, \ldots, n$.

By (5) we have $(E_{\delta_1} + \ldots + E_{\delta_n}) \subset D_\gamma$. Since $E_\gamma \cap D_\gamma = 0$, the conclusion follows immediately.

(7) $K = \sum_{\gamma \in \Gamma} \oplus E_\gamma$ and the sum is direct.

Let $D = \sum_{\gamma \in \Gamma} \oplus E_\gamma$; by (6) this sum is direct. Fix $\gamma \in \Gamma$, and let $G_\gamma = \sum_{\delta \neq \gamma} \oplus E_\delta$. Then $D = E_\gamma \oplus G_\gamma$; and $G_\gamma \subset D_\gamma$ by (5). To prove (7) it will be sufficient to prove that $G_\gamma = D_\gamma$. Suppose that $G_\gamma \neq D_\gamma$, and choose $x \in D_\gamma - G_\gamma$. By (4) there exists an element $s_\gamma \in R - M_\gamma$ such that $s_\gamma x = 0$. A fortiori, $s_\gamma x \in G_\gamma$.

Let δ be any index different from γ. Then $x = x_\delta + y_\delta$, where $x_\delta \in E_\delta$ and $y_\delta \in D_\delta$. By (4) there exists an element $s_\delta \in R - M_\delta$ such that $s_\delta y_\delta = 0$. Thus $s_\delta x = s_\delta x_\delta \in E_\delta \subset G_\gamma$. Let $I = \{r \in R \mid rx \in G_\gamma\}$, an ideal of R. We have shown that I is not contained in any maximal ideal of R. Thus $I = R$. But then $x \in G_\gamma$. This contradiction shows that $G_\gamma = D_\gamma$.

(8) R is an h-local ring.

By (3) we have $E_\gamma \cong K_{M_\gamma}$, and by (7) we have $K = \sum_\gamma \oplus E_\gamma$. Thus $K \cong \sum_\gamma \oplus K_{M_\gamma}$, and hence R is an h-local ring by Theorem 22. This completes the proof of the theorem.

Because a reflexive ring is an h-local ring and universal injectives localize well for h-local rings, we can manage to prove the following two important theorems on change of rings.

THEOREM 31. If R is a reflexive ring, then R_M is a reflexive ring for every maximal ideal M of R.

Proof. By (2) of the proof of Theorem 30 the injective envelope $E(R/M)$ of R/M is an R_M-module. By Theorem A3 it is an injective R_M-module. Since $R/M \cong R_M/M_{R_M}$, it is easy to see that $E(R/M)$ is the injective envelope of R_M/MR_M over R_M. By (3) of the proof of Theorem 30 we have $K_M \cong E(R/M)$, and the isomorphism is an R_M-isomorphism. Hence by Theorem 29, R_M is a reflexive ring.

THEOREM 32. Let R be an h-local ring. Then R is a reflexive ring if and only if R_M is a reflexive ring for every maximal ideal M of R.

Proof. One half of the theorem is proved by Theorem 31. Hence assume that R_M is a reflexive ring for all M. Then $K_M \cong E(R/M)$ by Theorem 29. Therefore K is an injective R-module by Theorem 24. Since $K \cong \sum_M \oplus K_M$ by Theorem 22, it follows from Theorem 29 that R is a reflexive ring.

5

NOETHERIAN REFLEXIVE RINGS

This chapter begins with an important property of modules generated by two elements.

THEOREM 33. Let R be an integral domain and A a torsion-free R-module that can be generated by two elements. Then A is a reflexive module if and only if $\text{Ext}_R^1(A, R) = 0$.

Proof. If rank $A = 2$, then the two generators of A must be free, and so $\text{Ext}_R^1(A, R) = 0$ and A is reflexive. Hence we can assume without loss of generality that $A = I$, a non-zero ideal of R. Let a and b be the generators of I and let $F = R \oplus R$. We now define:

(1) $\nu : F \to I$ by $\nu(r, s) = ra + sb$, for $r, s \in R$.

(2) $\eta : I^{-1} \to F$ by $\eta(x) = (xb, -xa)$, for $x \in I^{-1}$.

(3) $\lambda_1 : I^{-1} \to \text{Hom}_R(I, R)$ by $[\lambda_1(x)](c) = -xc$, for $x \in I^{-1}$ and $c \in I$.

(4) $\lambda_2 : F \to \text{Hom}_R(F, R)$ by $[\lambda_2(r, s)](u, v) = su - rv$, for $r, s, u, v \in R$.

(5) $\lambda_3 : I \to \text{Hom}_R(I^{-1}, R)$ by $[\lambda_3(c)](x) = cx$, for $c \in I$ and $x \in I^{-1}$.

It is easy to verify that we have a commutative diagram with exact rows:

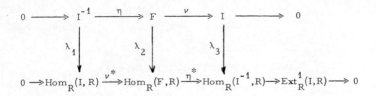

It is not hard to see that λ_1 and λ_2 are isomorphisms. From this it follows that η^* is an epimorphism if and only if λ_3 is an isomorphism. Since λ_3 can be identified with the inclusion $I \subset I^{-1-1}$, we have $\text{Ext}^1_R I, R) = 0$ if and only if I is a reflexive R-module.

THEOREM 34. Let M be a maximal ideal of R such that $M^{-1} \neq R$. Then M^{-1} can be generated by two elements if and only if $M^{-1}/R \cong R/M$.

Proof. If $M^{-1}/R \cong R/M$, take $u \in M^{-1} - R$. Then M^{-1} can be generated by 1 and u.

Conversely, suppose that M^{-1} can be generated by two elements u and v.

Case I: $MM^{-1} = R$.

In this case there exist elements $m_1, m_2 \in M$ such that $m_1 u + m_2 v = 1$. Since $m_1 u, m_2 v$ are in R, we can assume that $m_1 u \notin M$. Then $M + Rm_1 u = R$, and hence there exist elements $m_3 \in M$ and $r \in R$ such that $m_3 + rm_1 u = 1$. If $x \in M^{-1}$, then we have $x = m_3 x + (rm_1 x)u$. Since $m_3 x$ and $rm_1 x$ are in R, it follows that 1 and u generate M^{-1}. Therefore, $M^{-1}/R \cong R/M$.

Case II: $MM^{-1} \neq R$.

In this case we have $MM^{-1} = M$, and thus M^{-1}/M is a 2-dimensional vector space over R/M. Since M^{-1}/M maps onto

M^{-1}/R with kernel R/M, we see that M^{-1}/R is a 1-dimensional vector space over R/M. Therefore $M^{-1}/R \cong R/M$.

Statement (2) of the next theorem is a technical lemma that is needed for Theorem 40, while statement (3) repeats the two note theme.

THEOREM 35. Let R be an integral domain such that every ideal of R is reflexive. Then:

(1) The correspondence $A \to A^{-1}$ is a lattice anti-isomorphism of the non-zero fractionary ideals of R. In particular, if A and B are non-zero fractionary ideals, then
$(A \cap B)^{-1} = A^{-1} + B^{-1}$, and $(A + B)^{-1} = A^{-1} \cap B^{-1}$.

(2) Let a, b be elements of R with $b \neq 0$, and let $x = \frac{a}{b} + R$ in K. Define $(Rb : Ra) = \{r \epsilon R \mid ra \epsilon Rb\}$, and suppose that I is a proper ideal of R containing $(Rb : Ra)$. Then $Rx \cap (I^{-1}/R) \neq 0$.

(3) If M is a maximal ideal of R, then $M^{-1}/R \cong R/M$. Thus M^{-1} can be generated by two elements.

Proof. The proof of (1) is straightforward and is left to the reader.

(2). If $q = a/b$ then $Rx = R(q + R)$. Suppose that $Rx \cap (I^{-1}/R) = 0$. Then $(Rq + R) \cap I^{-1} = R$. Hence by statement (1) we have $R = (Rq + R)^{-1} + I$. But by (1) we have $(Rq + R)^{-1} = Rq^{-1} \cap R = (Rb : Ra)$. Therefore, $R = (Rb : Ra) + I = I$. This contradiction shows that $Rx \cap (I^{-1}/R) \neq 0$.

(3) By (1) there is a one-to-one correspondence between the set of ideals between M and R and the set of fractionary ideals between M^{-1} and R. Hence M^{-1}/R is a simple R-module and thus

$M^{-1}/R \cong R/M$. Therefore, by Theorem 34, M^{-1} can be generated by two elements.

The next theorem proves that in a Noetherian domain there are many reflexive prime ideals. The fact that they are of the form indicated will be quite useful.

THEOREM 36. Let R be a Noetherian domain and P a nonzero prime ideal of R. Then the following are equivalent:

(1) There exist elements a and b in R such that $P = (Rb:Ra)$.

(2) P is a reflexive ideal.

(3) P belongs to a principal ideal of R.

<u>Proof</u>. $(1) \Rightarrow (2)$. If $a, b \in R$ with $b \neq 0$, and if $x = a/b$, then $(Rb:Ra) = (Rx + R)^{-1}$. Hence $(Rb:Ra)$ is a reflexive ideal of R by Theorem 28.

$(2) \Rightarrow (3)$. Suppose that P is a reflexive ideal of R. Since R is a Noetherian ring, it follows that $R_P P$ is a reflexive ideal of R_P. Now P belongs to a principal ideal of R if and only if $R_P P$ belongs to a principal ideal of R_P. Thus we can assume that R is a Noetherian local ring with maximal ideal P. Since $P^{-1} \neq R$, we can find an element $x = a/b \in P^{-1} - R$, where $a, b \in R$. Then $Pa \subset Rb$ and $a \notin Rb$. Since P is a maximal ideal, it follows that P belongs to Rb.

$(3) \Rightarrow (1)$. Suppose that P belongs to a principal ideal Rb of R. We have a normal decomposition: $Rb = \bigcap_{i=1}^{m} Q_i$, where Q_i is P_i-primary for a prime ideal P_i and $P_1 = P$. Choose $c \in \bigcap_{i=2}^{m} Q_i$

such that $c \notin Q_1$. Then $(Q_1 : Rc)$ is P-primary, and so there exists

an integer $n > 0$ such that $P^n \subseteq (Q_1 : Rc)$ and $P^{n-1} \not\subseteq (Q_1 : Rc)$.

Choose $d \in P^{n-1}$ such that $d \notin (Q_1 : Rc)$ and let $a = cd$. Then we have

$P = (Rb : Ra)$.

DEFINITION. Let P be a prime ideal of a commutative ring

R. If there exists a chain of prime ideals of R, $P = P_0 \supsetneq P_1 \supsetneq \cdots \supsetneq P_n$

and no chain descending from P has more distinct prime ideals, the

rank of P is said to be n. The Krull dimension of R is defined to be

sup rank P, where P ranges over all the prime ideals of R.

The following theorem is a corollary to Theorem 36.

THEOREM 37. Let R be a Noetherian integral domain. Then

every prime ideal of rank 1 is reflexive.

Proof. A rank 1 prime ideal P of R belongs to every non-

zero element that it contains. Hence P is a reflexive ideal by

Theorem 36.

DEFINITIONS. An ideal I of a ring R is said to be irre-

ducible if whenever $I = J \cap L$, where J and L are ideals of R,

then $I = J$ or $I = L$.

If A is an R-module, we denote the injective envelope of A

by $E(A)$.

There is probably no theorem in the theory of injective modules

that is more useful than the one we will now prove. We will have occa-

sion to refer to it many times. Since a domain is reflexive if and only if

K is a universal injective, the importance of this theorem for reflexive

rings is predictable.

THEOREM 38. An R-module E is an indecomposable injective R-module if and only if $E \cong E(R/I)$, where I is an irreducible ideal of R. If E is an indecomposable injective, then for every non-zero $x \in E$, $O(x) = \{r \in R \mid rx = 0\}$ is an irreducible ideal of R and $E \cong E(R/O(x))$.

Proof. Let I be an irreducible ideal of R and let $E = E(R/I)$. Let A and B be non-zero submodules of E. Then $A \cap R/I = J/I$, where J is an ideal of R containing I. Since E is an essential extension of R/I, we have $J \neq I$. Similarly, $B \cap R/I = L/I$ where $L \supsetneq I$, and L is an ideal of R. Since $L \cap J \neq I$, we have $A \cap B \neq 0$. Thus E is an indecomposable R-module.

Conversely, let E be an indecomposable injective R-module and $x \in E$, $x \neq 0$. Let $E(Rx)$ be the injective envelope of Rx in E. Since $E(Rx)$ is a direct summand of E, we have $E = E(Rx)$. If $O(x) = I$, then $Rx \cong R/I$; and so $E \cong E(R/I)$. All that remains to be proved is that I is an irreducible ideal of R.

Suppose that $I = J \cap L$, where J and L are ideals of R such that $J \neq I$ and $L \neq I$. Let $E_1 = E(R/J)$ and $E_2 = E(R/L)$, and choose $x_1 \in E_1$ and $x_2 \in E_2$ such that $O(x_1) = J$ and $O(x_2) = L$. We define a monomorphism $f: Rx \to E_1 \oplus E_2$ by $f(rx) = (rx_1, rx_2)$ for $r \in R$. Then f extends to a homomorphism $g: E \to E_1 \oplus E_2$, since $E_1 \oplus E_2$ is injective. Since E is an essential extension of Rx, g is a monomorphism. We let $D = \text{Im } g$, and we have $D \cong E$.

Let $y = (x_1, x_2) \in D$; then D is an injective envelope of Ry in $E_1 \oplus E_2$ and $O(y) = I$. Let $r \in J - I$ and $s \in L - I$. Then $ry = (0, rx_2) \neq 0$ and $sy = (sx_1, 0) \neq 0$; and these are elements of D

such that $Rry \cap Rsy = 0$. But D is an indecomposable injective module, and hence $D = E(Rry)$ is an essential extension of Rry. Thus $Rry \cap Rsy \neq 0$. This contradiction shows that I is an irreducible ideal of R.

We now prove a lemma for Theorem 40.

THEOREM 39. Let M be a maximal ideal of R such that M^{-1} can be generated by two elements, and let I be a non-zero reflexive ideal of R such that $\text{Ext}_R^1(I, R) = 0$. If J is an ideal of R such that $J \supset I$ and $J/I \cong R/M$, then J is a reflexive ideal and $\text{Ext}_R^1(J, R) = 0$.

Proof. Let $S = R/M$. Then we have an exact sequence:

$$0 \to I \to J \to S \to 0.$$

Since $\text{Ext}_R^1(I, R) = 0$, we have a derived exact sequence:

$$0 \to J^{-1} \to I^{-1} \xrightarrow{\partial} \text{Ext}_R^1(S, R) \to \text{Ext}_R^1(J, R) \to 0.$$

If $\text{Im } \partial = 0$, then $J^{-1} = I^{-1}$, and we have $I = I^{-1-1} = J^{-1-1} \supset J$ and this contradicts $J/I \cong R/M$. Thus $\text{Im } \partial \neq 0$. Now $\text{Ext}_R^1(S, R) \cong \text{Hom}_R(R/M, K) \cong M^{-1}/R$. Since $\text{Im } \partial \neq 0$, we have $M^{-1}/R \neq 0$. Thus by Theorem 34, $M^{-1}/R \cong R/M$. Therefore ∂ is an epimorphism; and thus $\text{Ext}_R^1(J, R) = 0$.

If we apply $\text{Hom}_R(\cdot, R)$ to the preceding exact sequence and take into account that I is reflexive and that $\text{Ext}_R^1(S, R) \cong S$, we obtain an exact sequence:

$$0 \to I \to J^{-1-1} \xrightarrow{\beta} \text{Ext}_R^1(S, R) \ .$$

Thus again we have a situation where either $\beta = 0$ or β is onto. If $\beta = 0$, then $I \subset J \subset J^{-1-1} = I$ and this is a contradiction. Thus β is onto and we have $J^{-1-1}/I \cong S$. But $J/I \subset J^{-1-1}/I$, and therefore $J = J^{-1-1}$.

Remarks. There are examples to show that every ideal of R can be reflexive without R being a reflexive ring. There are also examples to show that K can be injective without being a universal injective. However, the next theorem shows that for Noetherian rings neither of these things can happen. Compare this theorem with Theorem 29!

THEOREM 40. Let R be a Noetherian integral domain. Then the following statements are equivalent:

(1) R is a reflexive ring.

(2) Every ideal of R is reflexive.

(3) R has Krull dimension 1, and M^{-1} can be generated by two elements for every maximal ideal M of R.

(4) K is injective.

Proof. (1) $\Rightarrow$ (2). Trivial.

(2) $\Rightarrow$ (3). Let P be a nonzero prime ideal of R. By Theorem 36 there exist elements a and b of R such that $P = (Rb:Ra)$. Since $P \neq 0$, we have $b \neq 0$, and we let $q = a/b$. Since $P \neq R$, $x = q + R$ is a nonzero element of K. Let M be a maximal ideal of R containing P. By Theorem 32(2), $Rx \cap (M^{-1}/R) \neq 0$. Hence there exists an element $r \in R$ such that $rq \in M^{-1} - R$. Since $rq \notin R$, we have $r \notin P$. However, $Mr \subset (Rb:Ra) = P$, and thus $M = P$. This shows that every nonzero prime ideal of R is maximal; that is R has Krull dimension 1.

If M is a maximal ideal of R, then by Theorem 35(3), M^{-1} can be generated by two elements.

(3) $\Rightarrow$ (4). Let I be a nonzero ideal of R, and choose a non-zero element $a \in I$. Since R is a Noetherian domain of Krull dimension one, I/Ra has finite length as an R-module. Thus we can find a chain of ideals of R:

$$I_0 = Ra \subset I_1 \subset I_2 \subset \ldots \subset I_n = I$$

such that I_{k+1}/I_k is a simple R-module for $k = 0, \ldots, n-1$. Now Ra is a reflexive ideal of R and $\operatorname{Ext}_R^1(Ra, R) = 0$. By assumption the hypotheses of Theorem 39 are satisfied. Thus I_1 is a reflexive ideal and $\operatorname{Ext}_R^1(I_1, R) = 0$. Continuing on up the chain we see finally that $\operatorname{Ext}_R^1(I, R) = 0$. Therefore, $\operatorname{Ext}_R^2(R/I, R) \cong \operatorname{Ext}_R^1(I, R) = 0$ for every ideal I of R. Thus $\operatorname{inj.dim}_R R = 1$; that is, K is injective.

(4) $\Rightarrow$ (1). Let P be a rank 1 prime ideal of R. By Theorem 37, P is a reflexive ideal of R, and hence $P^{-1}/R \neq 0$. It will be sufficient to prove that P is a maximal ideal of R. For then we will have $M^{-1}/R \neq 0$ for every maximal ideal M of R, proving that K is a universal injective R-module. Hence R will be a reflexive ring by Theorem 29.

By Theorem 36 there exist elements $a, b \in R$ with $b \neq 0$ such that $P = (Rb: Ra)$. Let $x = a/b + R$ in K, and let $E = E(Rx)$ be an injective envelope of Rx in K. Let $A = \{y \in E \mid Py = 0\}$. Since $Px = 0$, A is a nonzero R/P-module. We will prove that A is a torsion torsion-free R/P-module.

Suppose $y \in A$, $y \neq 0$, and let $O(y) = \{r \in R \mid ry = 0\}$. There exists $s \in R$ such that $sy \in Rx$ and $sy \neq 0$. Since $Rx \cong R/P$, and P

is a prime ideal of R, we see that $O(sy) = P$. Because $O(y) \subset O(sy)$ = P, it follows that A is a torsion-free R/P-module.

Now $A \cong \text{Hom}_R(R/P, E)$ is an injective R/P-module by Theorem A1. Since A is a torsion-free R/P-module, it is therefore a direct sum of copies of the quotient field of R/P. But $A = P^{-1}/R \cap E$, and hence A is a finitely generated R/P-module. This implies that R/P is a field, and thus P is a maximal ideal of R.

We have a further development of the theme of modules with two generators in the next theorem. It is still an open question whether or not the converse of this theorem is true. We will encounter a much more general result in Theorem 57.

THEOREM 41. Let R be a Noetherian domain of Krull dimension 1. If every maximal ideal of R can be generated by two elements, then R is a reflexive ring.

Proof. Let M be a maximal ideal of R. Since M is a rank 1 prime ideal of R, M is a reflexive ideal by Theorem 37. Because M can be generated by two elements we have $\text{Ext}_R^1(M, R) = 0$ by Theorem 33. Since $\text{Ext}_R^2(R/M, R) \cong \text{Ext}_R^1(M, R)$, we see that $\text{Ext}_R^2(R/M, R) = 0$ for every maximal ideal M of R. Let L be an R-module of finite length; that is, L has a composition series. Then an easy induction on the length of L shows that $\text{Ext}_R^2(L, R) = 0$.

Let I be a nonzero ideal of R. Since R has Krull dimension 1 and is Noetherian, R/I has finite length. Thus by the preceding paragraph we see that $\text{Ext}_R^2(R/I, R) = 0$. This implies that R has injective dimension 1; that is, that K is injective. Therefore, R is a reflexive ring by Theorem 40.

6

TORSIONLESS RINGS

DEFINITION. An integral domain R is said to be a <u>torsion-less ring</u>, if every reduced, torsion-free R-module of finite rank is torsionless.

The concept of a torsionless ring is a valuable aid in the study of reflexive rings, D-rings, and rings with certain other properties. In Theorem 42 we will prove that a domain is a torsionless ring if and only if it is complete and every proper submodule of its quotient field is a fractionary ideal. It follows immediately from this theorem that a valuation ring is a torsionless ring if and only if it is complete. Theorem 43 shows that the property of being a torsionless ring is an hereditary property in the sense defined in the introduction to these notes. In line with the thinking expressed there we make the conjecture, as yet open, that the integral closure of a torsionless ring is a Prüfer ring.

Theorem 44 is a lemma for Theorem 45 and for Theorem 63. However, the theorem is of great interest in itself -- it has as an immediate corollary the well-known theorem that the integral closure of a complete, Noetherian, local domain of Krull dimension 1 is a complete, discrete valuation ring which is finitely generated as a module over the domain. Finally, Theorem 45 characterizes Noetherian torsionless rings as precisely the complete, Noetherian, local domains of Krull dimension 1.

THEOREM 42. The following statements are equivalent:

(1) R is a torsionless ring.

(2) Every reduced, torsion-free R-module of rank ≤ 2 is torsionless.

(3) R is complete in the R-topology and every reduced torsion-free R-module of rank 1 is torsionless.

If R is a torsionless ring, then every reduced, torsion-free R-module of finite rank is complete in the R-topology.

<u>Proof</u>. (1) $\Rightarrow$ (2). Trivial.

(2) $\Rightarrow$ (3). Consider an exact sequence of the form

$$0 \to R \to A \xrightarrow{\;f\;} Q \to 0 .$$

Then A is a torsion-free R-module of rank two. Since $Q' = 0$, we have an exact sequence $0 \to A' \to R'$. Thus rank $A' \leq$ rank $R' = 1$, and hence by Theorem 27, A is not torsionless. It follows from (2) that A is not reduced. Thus A has a submodule B isomorphic to Q. Then either $B \subset \text{Ker } f$, or f is an isomorphism on B. Since Ker $f \cong R$, f is an isomorphism on B. Therefore $A = \text{Ker } f + B$ and Ker $f \cap B = 0$. This shows that the exact sequence splits, and we have proved that $\text{Ext}^1_R(Q, R) = 0$. By Theorem 9, R is complete in the R-topology.

(3) $\Rightarrow$ (1). Let A be a reduced, torsion-free R-module of rank n. We will prove that A is complete and torsionless by induction on n. If $n = 1$, then by assumption A is torsionless, and hence iso-

morphic to an ideal of R. Since R is complete, A is also complete by Theorem 14.

Assume that $n > 1$, and the assertion true for $n-1$. Let B be a pure, torsion-free submodule of A of rank $n-1$. Then A/B is torsion-free of rank 1. If A/B is not reduced, then $A/B \cong Q$, and by induction $\text{Ext}_R^1(A/B, B) = 0$. Therefore, $A \cong B \oplus A/B$ and A is not reduced. Hence A/B is torsionless, and thus $A/B \cong I$, a non-zero ideal of R.

Thus we have an exact sequence:

$$0 \to I' \to A' \to B' \to \text{Ext}_R^1(I, R).$$

Since $\text{Ext}_R^1(I, R) \cong \text{Ext}_R^2(R/I, R)$ it follows that $\text{Ext}_R^1(I, R)$ is torsion of bounded order. Thus rank $A' = $ rank $B' + $ rank I'. Now B and I are torsionless; hence by Theorem 27, rank $B' = n-1$ and rank $I' = 1$. Therefore rank $A' = n$ and by Theorem 27, A is torsionless.

We have an exact sequence:

$$\text{Ext}_R^1(Q, B) \to \text{Ext}_R^1(Q, A) \to \text{Ext}_R^1(Q, I) .$$

Since the ends of this sequence are zero by induction, we have $\text{Ext}_R^1(Q, A) = 0$. Thus by Theorem 9, A is complete.

THEOREM 43. Let R be a torsionless ring and S a ring extension of R in Q $(S \neq Q)$. Then S is also torsion-free.

Proof. By Theorem 43, S is a complete R-module. But then by Theorem 15, S is a complete S-module. Let A be a reduced, torsion-free S-module of rank 1. Then $\text{Hom}_R(A, R) \neq 0$, and $\text{Hom}_R(A, R) \subseteq \text{Hom}_S(A, S)$. Hence A is a torsionless S-module, and thus S is a torsionless ring by Theorem 42.

THEOREM 44. The following statements are equivalent:

(1) Every torsion-free R-module of finite rank is a direct sum of a finitely generated R-module and a divisible R-module.

(2) R is a complete, Noetherian, local domain of Krull dimension 1.

Proof. (1) $\Rightarrow$ (2). Since an ideal of R is reduced, it must be finitely generated. Thus R is a Noetherian ring. Let P be a non zero prime ideal of R. Then R_P is reduced, and hence R_P is a finitely generated R-module. But this forces R to be equal to R_P. Hence R is a Noetherian local domain of Krull dimension 1 with maximal ideal M.

Since a reduced, torsion-free R-module of finite rank is finitely generated, it follows from Theorem 27 that it is torsionless. Thus R is a torsionless ring. Hence by Theorem 42 R is complete in the R-topology. Now every non-zero ideal of R is M-primary, and hence the R-topology and the M-adic topology on R coincide. Therefore, R is complete in the M-adic topology.

(2) $\Rightarrow$ (1). Let A be an R-submodule of Q, $A \neq Q$. We will prove that A is a finitely generated R-module. Suppose that A is not finitely generated. We can assume that $R \subset A$, and we let $D = A/R$. Choose $a \in M$, $a \neq 0$, where M is the maximal ideal of R; then we have a descending chain of submodule $D \supset aD \supset a^2D \supset \ldots$. Since A is not isomorphic to an ideal of R, we see that $A^nD \neq 0$ for every integer $n > 0$.

Since R is a Noetherian ring of Krull dimension 1, it is easily seen that $M^{-1}/R \neq 0$ and that K is an essential extension of M^{-1}/R.

Thus every non-zero submodule of K contains a non-zero submodule of M^{-1}/R. Since M^{-1}/R is a finite-dimensional vector space over R/M, it follows that any descending chain of non-zero submodules of K has a non-zero intersection. Therefore, we have $\bigcap_n a^n D \neq 0$.

If S is the multiplicatively closed subset consisting of the powers of a, then R_S has Krull dimension 0, and hence $R_S = Q$. Therefore, an R-module is divisible if and only if it is divisible by the element a. Since $\bigcap_n a^n A$ is divisible by a, and is a proper R-submodule of Q, we have $\bigcap_n a^n A = 0$. Thus if we let $I_n = a^n A \cap R$, we have $I_1 \supset I_2 \supset I_3 \supset \ldots$ and $\bigcap_n I_n = 0$. Since R is a complete, Noetherian, local ring, it is not hard to show that given any integer $t > 0$ there is an integer $n > 0$ such that $I_n \subset M^t$. On the other hand, every ideal I_n is M-primary since R has Krull dimension 1; and thus given any integer $n > 0$ there is an integer $m > 0$ such that $M^m \subset I_n$. These remarks show that the M-adic and the I_n topologies on R are the same. Thus R is complete in the I_n topology.

Choose an element $x \in A$ which maps onto a non-zero element of $\bigcap_n a^n D$. Then $x \notin R$; but for each integer $n > 0$ we have $x = a^n x_n + r_n$, where $x_n \in A$ and $r_n \in R$. Clearly $\{r_n\}$ is a Cauchy sequence in R in the I_n-topology; and hence r_n converges to an element $r \in R$. But then $x - r \in \bigcap_n a^n A = 0$, and thus $x \in R$. This contradiction shows that A is a finitely generated R-module.

Let B be a reduced, torsion-free R-module of finite rank. We will prove that B is finitely generated by induction on n, the case $n = 1$ being the case just considered. Assume $n > 1$ and let C be a pure submodule of B of rank $n-1$. Then B/C is torsion-free of rank 1, and

C is finitely generated by induction. If B/C is reduced, then it is finitely generated, and consequently B is also finitely generated.

Hence assume that B/C is not reduced; then B/C $\cong$ Q. Since R is complete, and reduced torsion-free rank 1 modules are finitely generated, R is a torsionless ring by Theorem 42. Therefore C is complete by Theorem 42, and hence $\text{Ext}_R^1(B/C, C) = 0$. Thus B = C $\oplus$ B/C, and B is not reduced. This contradiction shows that B is finitely generated and proves the theorem.

THEOREM 45. Let R be a Noetherian integral domain. Then the following statements are equivalent:

(1) R is a torsionless ring.

(2) $\text{Ext}_R^1(Q, A) = 0$ for every torsion-free R-module A of rank 1; that is, every proper submodule of Q is cotorsion.

(3) R is a complete, Notherian, local domain of Krull dimension 1.

Proof. (1) $\Rightarrow$ (2). This is an immediate consequence of Theorem 42.

(2) $\Rightarrow$ (3). We will prove first that R has only one prime ideal of rank 1. This will prove that R is a local domain of Krull dimension 1, for by Krull's principal ideal theorem every non-zero element of R that is not a unit is an element of a rank 1 prime ideal.

Suppose that R has two distinct prime ideals P_1 and P_2 of rank 1. Let S = R - ($P_1 \cup P_2$); then R_S is a Noetherian domain of Krull dimension 1 with only two distinct maximal ideals. Thus R_S is an h-local ring; and by Theorem 22, Q/R_S has a non-trivial direct sum decomposition. But by (2), R_S is a complete R-module, and hence by

Theorem 16, Q/R_S is indecomposable. This contradiction shows that R is a Noetherian, local domain of Krull dimension 1.

Let M be the maximal ideal of R. Then the R-topology and the M-adic topology on R are the same. Since R is complete in the R-topology by (2), it is therefore also complete in the M-adic topology.

(3) $\Rightarrow$ (1). By Theorem 44 every reduced, torsion-free R-module of finite rank is finitely generated. It then follows from Theorem 27 that R is a torsionless ring.

7

COMPLETELY REFLEXIVE RINGS

DEFINITION. R is said to be a <u>completely reflexive ring</u> if every reduced, torsion-free R-module of finite rank is reflexive. Clearly a completely reflexive ring is both a reflexive ring and a torsionless ring. The next theorem provides a justification for the terminology.

THEOREM 46. The following statements are equivalent:

(1) R is a completely reflexive ring.

(2) R is both a reflexive and a torsionless ring.

(3) R is a reflexive ring that is complete in the R-topology.

<u>Proof.</u> (1) $\Rightarrow$ (2). This is an obvious statement.

(2) $\Rightarrow$ (3). Theorem 42 states that R is complete in the R-topology.

(3) $\Rightarrow$ (1). By Theorem 42 it will be sufficient to prove that if A is a reduced, torsion-free R-module of rank 1, then $A' \neq 0$. Since R is complete, we see that $R \cong \mathrm{Hom}_R(K, K)$ by Theorem 10. Using this fact and Theorem A1 we have $A' = \mathrm{Hom}_R(A, R) \cong \mathrm{Hom}_R(A, \mathrm{Hom}_R(K, K)) \cong \mathrm{Hom}_R(K \otimes_R A, K)$. Now $K \otimes_R A \neq 0$, since A is reduced; and K is a universal injective by Theorem 29. Hence $\mathrm{Hom}_R(K \otimes_R A, K) \neq 0$; that is, $A' \neq 0$.

DEFINITION. If B is an R-module, we will denote $\text{Hom}_R(B, K)$ by B^*. Thus we have a canonical map $\phi: B \to B^{**}$ given by

$$[\phi(x)](f) = f(x)$$

for all $x \in B$ and $f \in B^*$. This definition is entirely similar to that of $\lambda: B \to B''$, except that here we are considering duality with respect to K instead of with respect to R. If I is a non-zero ideal of R, then $(R/I)^*$ is isomorphic to the R-submodule consisting of all $x \in R$ such that $Ix = 0$; i.e. $(R/I)^* \cong I^{-1}/R$. If f is an R-homomorphism from I^{-1}/R into K, then $I(\text{Im} f) = 0$, and hence $\text{Im} f \subset I^{-1}/R$. Thus $(R/I)^{**}$ is isomorphic to the endomorphism ring $\text{Hom}_R(I^{-1}/R, I^{-1}/R)$.

The next theorem gives us a very useful characterization of completely reflexive rings in terms of duality with respect to K.

THEOREM 47. The following statements are equivalent:

(1) R is a completely reflexive ring.

(2) $\phi: A \to A^{**}$ is an isomorphism for every R-module A which is a submodule of a finitely generated R-module.

Proof. (1) $\Longrightarrow$ (2). Let A be a cyclic R-module. If $A \cong R$, then ϕ is the canonical embedding $R \to \text{Hom}_R(K, K)$. By Theorem 46 R is complete and thus ϕ is an isomorphism. Hence assume that $A \cong R/I$, where I is a non-zero ideal of R. The kernel of $\phi: R/I \to (R/I)^{**}$ is I^{-1-1}/I, which is zero, since every ideal of R is reflexive. Now $(R/I)^* \cong I^{-1}/R$, and hence we identify $(R/I)^{**}$ with $\text{Hom}_R(I^{-1}/R, K)$. Since K is injective by Theorem 29, every R-homomorphism $f: I^{-1}/R \to K$ can be extended to an R-homomorphism from K into K. But $\text{Hom}_R(K, K) \cong R$, and thus f is multiplication by an element $r \in R$. It is readily verified that $\phi(r+I) = f$ and thus ϕ is an isomorphism. Therefore, $\phi: A \to A^{**}$ is an isomorphism for every cyclic R-module A.

Now let A be a finitely generated R-module and B a submodule of A with fewer generators such that $C = A/B$ is a cyclic R-module. Since K is injective, we have a commutative diagram with exact rows:

$$
\begin{array}{ccccccccc}
0 & \to & B & \to & A & \to & C & \to & 0 \\
& & \phi_B \downarrow & & \phi_A \downarrow & & \phi_C \downarrow & & \\
0 & \to & B^{**} & \to & A^{**} & \to & C^{**} & \to & 0
\end{array}
$$

We have proved that ϕ_C is an isomorphism, and ϕ_B is an isomorphism by induction on the number of generators. Thus ϕ_A is an isomorphism.

In general, if B is any submodule of a finitely generated R-module A, and $C = A/B$, then C is finitely generated. In this case ϕ_A and ϕ_C are isomorphisms by the preceding paragraph, and hence ϕ_B is also an isomorphism.

$(2) \Rightarrow (1)$. Let I be any non-zero ideal of R. Since the kernel of $\phi: R/I \to (R/I)^{**}$ is I^{-1-1}/I, we have $I = I^{-1-1}$. Thus every ideal of R is reflexive. We will prove next that K is injective, and for this it is sufficient to prove that $\mathrm{Ext}_R^1(I, R) = 0$. Since I is reflexive, there is an ideal J of R such that $I \cong J^{-1}$. Hence it is sufficient to prove that $\mathrm{Ext}_R^1(J^{-1}, R) = 0$.

Applying the functor $\mathrm{Hom}_R(J^{-1}/R, \cdot)$ to exact sequence (I) we obtain an isomorphism $\mathrm{Hom}_R(J^{-1}/R, K) \cong \mathrm{Ext}_R^1(J^{-1}/R, R)$. Since

$R/J \cong (R/J)^{**} \cong \operatorname{Hom}_R(J^{-1}/R, K)$, we see that $R/J \cong \operatorname{Ext}_R^1(J^{-1}/R, R)$.

Applying the functor $\operatorname{Ext}_R^1(\cdot, R)$ to the exact sequence:

$$0 \to R \to J^{-1} \to J^{-1}/R \to 0$$

we obtain an epimorphism $\operatorname{Ext}_R^1(J^{-1}/R, R) \to \operatorname{Ext}_R^1(J^{-1}, R) \to 0$. Thus $\operatorname{Ext}_R^1(J^{-1}, R)$ is a homomorphic image of R/J and hence is a cyclic torsion module. Since $*$ is a right exact functor, we obtain from the epimorphism: $\operatorname{Hom}_R(J^{-1}, K) \to \operatorname{Ext}_R^1(J^{-1}, R) \to 0$ a monomorphism $0 \to \operatorname{Ext}_R^1(J^{-1}, R)^* \to (J^{-1})^{**}$. But $(J^{-1})^{**} \cong J^{-1}$ and $\operatorname{Ext}_R^1(J^{-1}, R)^*$ is a torsion R-module. Therefore $\operatorname{Ext}_R^1(J^{-1}, R)^* = 0$. Since $\operatorname{Ext}_R^1(J^{-1}, R)$ is a cyclic R-module we have $\operatorname{Ext}_R^1(J^{-1}, R) \cong \operatorname{Ext}_R^1(J^{-1}, R)^{**} = 0$. Thus we have proved that K is injective.

By Theorem 29, R is a reflexive ring. Since $R \cong \operatorname{Hom}_R(K, K)$, R is complete in the R-topology. Hence by Theorem 46, R is a completely reflexive ring.

The next theorem describes many of the properties of a completely reflexive ring. It is an important lemma for Theorems 50 and 53 as well as for Theorem 72.

THEOREM 48. Let R be a completely reflexive ring and S a ring extension of R in Q, $(S \neq Q)$. Then:

(1) S is a complete, quasi-local ring. Thus R is quasi-local ring and the prime ideals of R (and of S) are linearly ordered.

(2) S is a completely reflexive ring if and only if S^{-1} is a principal ideal of S.

(3) If A is an S-submodule of Q, then Q/A is an injective S-module if and only if A^{-1} is a flat S-module. Furthermore, if N is the maximal ideal of S, then Q/S^{-1} is the injective envelope of S/N as an S-module.

(4) If M is the maximal ideal of R, and if $B \subset A$ are R-submodule of Q such that $M^k A \subset B$ for some integer k, then A/B has finite length (that is, A/B has a composition series).

Proof. Let A be a non-zero R-submodule of Q, $(A \neq Q)$. We will first establish a number of identities.

(a) $\qquad (K \otimes_R A)^* \cong A^{-1}$.

Since R is complete, we have $R \cong \operatorname{Hom}_R(K, K)$, and thus using Theorem A1, $A^{-1} = \operatorname{Hom}_R(A, R) \cong \operatorname{Hom}_R(A, \operatorname{Hom}_R(K, K)) \cong$ $\operatorname{Hom}_R(K \otimes_R A, K) = (K \otimes_R A)^*$.

The second identity is:

(b) $\qquad A^* \cong K \otimes_R A^{-1} \cong Q/A^{-1}$.

If we apply the exact functor $*$ to exact sequence II we have an exact sequence:

$$0 \to (K \otimes_R A)^* \to Q^* \to A^* \to 0 .$$

Since R is complete, $\operatorname{Ext}^1_R(Q, R) = 0$. Therefore applying the functor $\operatorname{Hom}_R(Q, \cdot)$ to exact sequence (I) we have the isomorphisms: $Q^* = \operatorname{Hom}_R(Q, K) \cong \operatorname{Hom}_R(Q, Q) \cong Q$. Since $(K \otimes_R A)^* \cong A^{-1}$ by (a), the above exact sequence becomes:

$$0 \to A^{-1} \to Q \to A^* \to 0 .$$

Applying the functor $K \otimes_R \cdot$ to this sequence we obtain (b).

The third identity is:

(c) $\quad K \otimes_R A \cong (K \otimes_R A)^{**}$.

If we apply (b) to A^{-1} we have $(A^{-1})^* \cong K \otimes_R A^{-1-1} \cong$
$K \otimes_R A$, since A is a reflexive module. Now $(A^{-1})^{**} \cong A^{-1}$ by
Theorem 47. Thus we have $(K \otimes_R A)^{**} \cong (A^{-1})^{***} \cong (A^{-1})^* \cong$
$K \otimes_R A$.

We are now ready to prove the theorem.

(1)   Since K is an injective R-module, we see by an easy application
of Theorem A1 that $\text{Hom}_R(S, K)$ is an injective S-module. By (b) we
have $S^* \cong Q/S^{-1}$, and of course $S^* = \text{Hom}_R(S, K)$. Thus Q/S^{-1} is an
injective S- module. Since S^{-1} is complete as an R-module, it is
complete as an S-module by Theorem 15. Therefore, Q/S^{-1} is an indecomposable injective S-module by Theorem 16.

Let N be a maximal ideal of S. Then using Theorem A1, we
have $\text{Hom}_S(S/N, S^*) = \text{Hom}_S(S/N, \text{Hom}_R(S, K)) \cong \text{Hom}_R(S/N \otimes_S S, K) \cong$
$\text{Hom}_R(S/N, K) = (S/N)^*$. Because K is an universal injective R-
module, we have $(S/N)^* \neq 0$. Thus S^* contains a copy of S/N. By
Theorem 38, S^* is the injective envelope over S of S/N. This proves
the second statement of (3).

Suppose that S has a second maximal ideal N_1. Then by the
preceding paragraph S^* is also the injective envelope over S of S/N_1.
Therefore S^* contains two non-isomorphic simple S-modules T_1 and
T_2. Hence $T_1 \cap T_2 = 0$, which contradicts Theorem 38. Therefore, S
is a quasi-local ring with maximal ideal N. Since S is complete in the
R-topology, it is complete in the S-topology by Theorem 15. This
proves (1).

(2)   Since S is complete in the S-topology, S is a completely reflexive ring if and only if S is a reflexive ring by Theorem 46. And by Theorem 29, S is a reflexive ring if and only if Q/S is the injective envelope over S of S/N. Thus by the second statement of (3), which we have already proved, S is a completely reflexive ring if and only if $Q/S \cong Q/S^{-1}$. Since S and S^{-1} are complete S-modules, we see that $Q/S \cong Q/S^{-1}$ if and only if $S \cong S^{-1}$ by the duality of Theorem 6. Thus S is a completely reflexive ring if and only if S^{-1} is a principal ideal of S.

(3)   Let A be an S-submodule of Q, $(A \neq Q)$, and let $C = K \otimes_R A$. Then by (c) we have $C \cong C^{**}$. Thus for an S-module B we have by Theorem A2, $\mathrm{Ext}_S^1(B, C) \cong \mathrm{Ext}_S^1(B, C^{**}) \cong (\mathrm{Tor}_1^S(B, C^*))^*$. Since K is a universal injective R-module, it follows that $\mathrm{Ext}_S^1(B, C) = 0$ if and only if $\mathrm{Tor}_1^S(B, C^*) = 0$. Since $C^* \cong A^{-1}$ by (a), and $C \cong Q/A$, it follows that Q/A is an injective S-module if and only if A^{-1} is a flat S-module.

(4)   Let $B \subset A$ be R-submodules of Q such that $M^k A \subset B$ for some integer k, where M is the maximal ideal of R. Then A/B has finite length as an R-module if and only if $(M^i A + B)/(M^{i+1} A + B)$ has finite length for every integer i. Thus without loss of generality we can assume that $MA \subset B$, and that $A \neq Q$. Since $A^{-1} \neq 0$, A is isomorphic to an ideal of R. Thus A/B is a submodule of a cyclic R-module. Hence by Theorem 47 we have $(A/B) \cong (A/B)^{**}$. Since A/B is annihilated by M, we have $(A/B)^* = \mathrm{Hom}_A(A/B, M^{-1}/R)$. But $M^{-1}R \cong R/M$ by Theorem 35. Thus $(A/B)^* \cong \mathrm{Hom}_R(A/B, R/M)$. Therefore, $(A/B) \cong (A/B)^{**}$ implies that A/B is a finite-dimensional vector space over R/M; that is, A/B has finite length.

This concludes the proof of the theorem.

The following theorem is due to I.S. Cohen. It provides an important criterion for determining when a ring is Noetherian. We will need it for Theorems 50, 57, and 100.

THEOREM 49. Let R be a commutative ring. Then R is a Noetherian ring if and only if every prime ideal of R is finitely generated.

Proof. Of course, if R is a Noetherian ring, then every prime ideal of R is finitely generated. Conversely, assume that every prime ideal of R is finitely generated. Suppose that R is not a Noetherian ring. Then there exist ideals which are not finitely generated; and hence, by Zorn's Lemma, an ideal P of R that is maximal with respect to this property. We will prove that P is a prime ideal of R; and this contradiction will establish the fact that R is Noetherian.

Suppose that P is not a prime ideal of R. Then there exist elements a and $b \in R - P$ such that $ab \in P$. Thus both $P + Rb$ and $(P:Rb)$ properly contain P. By the maximality of P both of these ideals are finitely generated. Let $a_1 + r_1 b, \ldots, a_n + r_n b$ generate $P + Rb$, where $a_i \in P$ and $r_i \in R$ for $i = 1, \ldots, n$; and let $c_1, \ldots, c_m$ generate $(P:Rb)$. It is then easy to verify that $a_1, \ldots, a_n$, $c_1 b, \ldots, c_m b$ generate P. This contradiction shows that R is a Noetherian ring.

The next theorem characterizes those completely reflexive rings that are Hausdorff in every ideal-adic topology.

THEOREM 50. The following statements are equivalent:

(1) R is a completely reflexive ring and $\bigcap\limits_{n=1}^{\infty} M^n = 0$, where M is the maximal ideal of R.

(2) R is a complete, Noetherian, local domain of Krull dimension 1 such that K is injective.

Proof. (1) $\Rightarrow$ (2). Let $A = \bigcup\limits_{n=1}^{\infty} (M^n)^{-1}$; since the modules $(M^n)^{-1}$ are linearly ordered, A is an R-submodule of Q. Clearly $A^{-1} \subset (M^n)^{-1-1} = M^n$ for all $n \geq 1$. Therefore, $A^{-1} = 0$. Since R is a torsionless ring, we see that $A = Q$.

Let $a \in M$, $a \neq 0$; then there exists an integer $n > 0$ such that $1/a \in (M^n)^{-1}$. Hence we have $M^n = (M^n)^{-1-1} \subset Ra$. This shows that M is the only non-zero prime ideal of R. Therefore, the Krull dimension of R is 1. Furthermore, since $M^n \subset Ra$, we see by Theorem 48 (4) that M/Ra has finite length. Thus M is finitely generated. By Theorem 49, R is a Noetherian ring. That R is a complete, Noetherian, local domain of Krull dimension 1 and that K is injective now follow from Theorem 46 and Theorem 29.

(2) $\Rightarrow$ (1) By Theorem 40, R is a reflexive ring. Since the M-adic topology and the R-topology are the same for 1-dimensional Noetherian local domains, R is complete in the R-topology. Thus R is a completely reflexive ring by Theorem 46. In every Noetherian domain we have $\bigcap\limits_{n=1}^{\infty} I^n = 0$ for every ideal I of R.

MAXIMAL VALUATION RINGS

DEFINITIONS. Let V be an integral domain with quotient field Q. V is said to be a valuation ring, if given any $x \in Q$, $x \neq 0$, then either x or x^{-1} is in V. It is not hard to see that a valuation ring V is a quasi-local ring, that the V-submodules of Q are linearly ordered, and that V is integrally closed in Q.

A valuation ring is said to be a discrete valuation ring, if it is a Noetherian ring. It is easy to show that if V is a valuation ring with maximal ideal N, then V is Noetherian if and only if $\bigcap_k N^k = 0$. If this happens, then N is a principal ideal and is the only non-zero prime ideal of V.

Suppose that R and S are quasi-local domains with the same quotient field Q, and with maximal ideals M and N, respectively. S is said to dominate R if $R \subset S$ and $N \cap R = M$. It is not difficult to prove that R is a valuation ring if and only if it is not dominated by any quasi-local subring of Q bigger than itself. In general, the set of quasi-local subrings of Q that dominate R form an inductive set. Hence by Zorn's Lemma this set has a maximal element, which is then necessarily a valuation ring. Thus R is dominated by a valuation ring.

It can be shown that a domain is integrally closed in its field of quotients if and only if it is the intersection of the valuation rings which contain it.

Suppose that A is a module over the commutative ring R. A is said to be <u>linearly compact</u> if every finitely solvable set of congruences:

$$x \equiv x_\alpha \pmod{A_\alpha}$$

(where $x_\alpha \in A$ and the A_α's are submodules of A) has a simultaneous solution in A.

A valuation ring V with quotient field Q is said to be a <u>maximal valuation ring</u>, if Q is a linearly compact V-module. It is easy to see that V is a maximal valuation ring if and only if V itself is a linearly compact V-module.

This is a cumbersome definition of a maximal valuation ring, and in the following theorem we will prove the equivalence of a number of more elegant and more useful definitions. We note that it follows from this theorem that a discrete valuation ring is a maximal valuation ring if and only if it is complete. We should also point out that a maximal valuation ring does not necessarily have any maximality property (in any sense of containment) within its quotient field. It is, however, maximal with respect to enlargements which preserve its value group and residue class field, and this is the origin of the name.

THEOREM 51. Let R be an integral domain with quotient field Q and $K = Q/R$. Then the following statements are equivalent:

(1) $\text{Ext}_R^1(A, S) = 0$ whenever A is a torsion-free R-module and S is a torsion-free R-module of finite rank.

(2) If B is a torsion-free R-module, and S is a pure sub-module of B of finite rank, then S is a direct summand of B.

(3) R is a complete integral domain and $\text{inj.dim}_R I = 1$ for every non-zero ideal I of R.

(4) R is a complete valuation ring and K is injective.

(5) R is a maximal valuation ring.

Proof. (1) $\Longleftrightarrow$ (2) Statement (2) is merely a rephrasing of statement (1) using the fact that $\text{Ext}_R^1(A, S)$ is the set of equivalence classes of extensions of S by A.

(1) $\Longrightarrow$ (3) Since $\text{Ext}_R^1(Q, R) = 0$, R is complete by Theorem 9. Let I and J be ideals of R. Then $\text{Ext}_R^2(R/J, I) \cong \text{Ext}_R^1(J, I)$; and $\text{Ext}_R^1(J, I) = 0$ by hypothesis. Therefore, $\text{inj.dim}_R I = 1$.

(3) $\Longrightarrow$ (4) Let a and b be two non-zero elements of R, and let $I = Ra \cap Rb$. Since $\text{inj.dim}_R I = 1$, we see that Q/I is an injective R-module. Since R is complete, I is complete by Theorem 14. Then Q/I is an indecomposable R-module by Theorem 16. Thus Q/I is an indecomposable injective R-module.

Let $x = 1+I \in Q/I$, and let $O(x) = \{ r \in R \mid rx = 0 \}$. Then $O(x) = I$, and by Theorem 38, I is an irreducible ideal. Hence either $a \in Rb$ or $b \in Ra$. Thus R is a valuation ring. R is complete and K is injective by assumption (3).

(4) $\Longrightarrow$ (5). Suppose that the following set of congruences is finitely solvable:

$$x \equiv x_\alpha \pmod{I_\alpha}$$

where $r_\alpha \in R$ and I_α is an ideal of R for every α. We must find a simultaneous solution for these congruences.

<u>Case I:</u> $\bigcap_{\alpha} I_{\alpha} = 0$.

For each non-zero ideal I of R we choose an index α such that $I_{\alpha} \subset I$ and define $r_I = r_{\alpha}$. We then define an element $\tilde{x}$ of $\prod (R/I)$ by $\tilde{x} = \langle r_I + I \rangle$. Since the congruences are finitely solvable, $\tilde{x}$ is independent of the choices we have made. Now in fact we have $\tilde{x} \in H$, the completion of R. For suppose that J is a non-zero ideal of R contained in the given ideal I. We have already chosen an index β such that $I_{\beta} \subset J$, and an element $r_J = r_{\beta}$. Since the congruences are finitely solvable, there is an element $t \in R$ such that $t - r_{\beta} \in I_{\beta}$ and $t - r_{\alpha} \in I_{\alpha}$. Thus we have $r_J - r_I = r_{\beta} - r_{\alpha} = (r_{\beta} - t) + (t - r_{\alpha}) \in I_{\beta} + I_{\alpha} \subset J + I = I$. Hence $\tilde{x} \in H$, by the definition of H. But R is complete, and hence $R \cong H$. Thus there exists an element $r \in R$ such that $\langle r + I \rangle = \tilde{x} = \langle r_I + I \rangle$. Therefore, $r - r_{\alpha} \in I_{\alpha}$ for every index α, and hence r is a solution of the congruences.

<u>Case II:</u> $\bigcap I_{\alpha} = I \neq 0$. (The proof of this case is due originally to I. Kaplansky. See [17, Th. 4]).

If $I_{\beta} = I$ for some β, then r_{β} is a solution of the congruences. Hence we can assume that $I_{\alpha} \neq I$ for all indices α. Take c in I, $c \neq 0$, and define $J = \{ b \in R \mid bI \subsetneq cR \}$. J is an ideal of R, and for every $b \in J$ there is an index α_b such that $bI_{\alpha_b} \subsetneq cR$. We define an R-homomorphism $f: J \to Q/cR$ by $f(b) = br_{\alpha_b} + cR$ for every $b \in J$. It is easily verified that this definition is independent of the choices we have made. Since Q/cR is isomorphic to K it is injective, and hence there is an element $q \in Q$ such that $f(b) = bq + cR$ for every $b \in J$. We will show that $q \in R$ and that q is a solution of the congruences.

Let α_0 be a fixed index and take $d \in I_{\alpha_0} - I$. Since $d \notin cR$, we have $c \in dR$. Hence there exists $y \in R$ such that $c = dy$. In fact $y \in J$. For if $y \notin J$, then $cR \subset yI$. Hence there exists $z \in I$ such that $c = yz$. Thus $yd = c = yz$, and so $d = z \in I$. This contradiction shows that $y \in J$.

Now there exists an index β such that $I_\beta \subset I_{\alpha_0}$ and such that $yI_\beta \subsetneq cR$. Then $yr_\beta + cR = f(y) = yq + cR$, and so $y(r_\beta - q) \in cR$. Hence there is an $r \in R$ such that $y(r_\beta - q) = cr = ydr$. Therefore, $r_\beta - q = dr \in I_{\alpha_0}$. Since $I_\beta \subset I_{\alpha_0}$, we have $r_{\alpha_0} - q = (r_{\alpha_0} - r_\beta) + (r_\beta - q) \in I_{\alpha_0}$. Since this is true for every index α_0, q is a solution of the congruences; and of course $q \in R$.

Thus we have solved the system of congruences in both cases, which proves that R is a maximal valuation ring.

$(5) \Rightarrow (1)$. Let A be a torsion-free R-module and S a torsion-free R-module of finite rank. Suppose that we have proved that $\text{Ext}_R^1(A, S) = 0$ when S has rank 1. Assume that S has rank $n > 1$, and let C be a pure submodule of S of rank $n-1$. Then S/C is torsion-free of rank one, and we have an exact sequence:

$$\text{Ext}_R^1(A, C) \to \text{Ext}_R^1(A, S) \to \text{Ext}_R^1(A, S/C) .$$

The first module is zero by induction on the rank of S, and the last module is zero by the rank 1 case we have assumed. Thus $\text{Ext}_R^1(A, S) = 0$. Hence it is sufficient to assume that rank $S = 1$. Thus we can assume that S is an R-submodule of Q. Since $\text{Ext}_R^1(A, Q) = 0$, we can assume that S is reduced.

We will prove first that S is a cotorsion R-module. By Theorem 9 this is equivalent to proving that S is complete in the

R-topology. Let $\widetilde{x} \in \widetilde{S}$, the completion of S. Then $\widetilde{x} = < x_r + rS >$, where $x_r \in S$ and r ranges over the non-zero elements of R. The system of congruences:

$$x \equiv x_r \pmod{rS}$$

is finitely solvable by the definition of $\widetilde{S}$ as $\varprojlim S/rS$. Since R is a maximal valuation ring, there is a simultaneous solution $y \in S$. Then $\widetilde{x} = < y + rS >$, and hence $\widetilde{S} \cong S$. Therefore S is a cotorsion R-module.

We prove next that inj. dim. $_R S = 1$; that is, we will prove that Q/S is an injective R-module. Let I be an ideal of R and $f: I \rightarrow Q/S$ be an R-homomorphism. Since Q/S is a divisible R-module, for each $a \in I$ there is an element $x_a \in Q$ such that $f(a) = a(x_a + S)$. For each $a \in I$ define $D_a = \{q \in Q \mid aq \in S\}$. Then the system of congruences:

$$x \equiv x_a \pmod{D_a}$$

is finitely solvable in Q. Since R is a maximal valuation ring, there is a simultaneous solution $y \in Q$. Define $g: R \rightarrow Q/S$ by $g(r) = r(y + S)$. Then g is an extension of f to R. This proves that Q/S is injective; that is, inj. dim. $_R S = 1$.

If we apply the functor $\text{Ext}_R^1(\cdot, S)$ to exact sequence (II) for A we obtain an exact sequence:

$$\text{Ext}_R^1(Q \otimes_R A, S) \rightarrow \text{Ext}_R^1(A, S) \rightarrow \text{Ext}_R^2(K \otimes_R A, S).$$

Since S is a cotorsion R-module, the first term of this sequence is zero; and since inj. dim. $_R S = 1$, the last term is zero. Therefore, $\text{Ext}_R^1(A, S) = 0$.

This concludes the proof of the theorem.

We follow Nagata's proof of the next theorem almost exactly [29, Th. 11.11].

THEOREM 52. (Independence of valuations): Let $V_1, \ldots, V_m$ be valuation rings with the same quotient field Q and assume that $V_i \not\subseteq V_j$ for any $i \neq j$. Let N_i be the maximal ideal of V_i, let R be the intersection of the V_i, and set $M_i = N_i \cap R$. Then we have:

(1) $R_{M_i} = V_i$ for all i.

(2) The M_i are the only maximal ideals of R and they are all distinct.

Proof. Let $a \in Q$ and define $u_n = 1 + a + a^2 + \ldots + a^{n-1}$ for each integer $n \geq 1$. We will prove first that there exists an integer $t \geq 1$ such that both $1/u_t$ and a/u_t are in V_i for every $i = 1, \ldots, m$.

If $a \notin V_i$, then $1/u_n$ and a/u_n are in V_i for all $n \geq 3$. For since $a \notin V_i$, we have $1/a \in V_i$. Since $u_{n-1} = (1/a)(u_n - 1)$, if $u_n \in V_i$, then $u_{n-1} \in V_i$. But $u_2 = 1 + a \notin V_i$, and thus we must have $u_n \notin V_i$ for all $n \geq 2$. Thus $1/u_n \in V_i$ for all $n \geq 1$. Now $u_n/a = (1/a) + u_{n-1}$ cannot be in V_i for all $n \geq 3$, and hence $a/u_n \in V_i$ for all $n \geq 3$.

Thus without loss of generality we can assume that $a \in V_i$ for every i, and we only need to prove that there exists an integer $t \geq 3$ such that $u_t \notin N_i$ for all i.

Let p_i be the characteristic of V_i/N_i if it is greater than 0; otherwise let $p_i = 1$. Let q_i be the smallest positive integer such that $a^{q_i} \equiv 1 \pmod{N_i}$ if such a positive integer exists; otherwise let $q_i = 1$. Let $s = \prod_{i=1}^m p_i q_i$ and let $t = 2s + 1$. If $a \equiv 1 \pmod{N_i}$, then

$u_n \equiv n \cdot 1 \pmod{N_i}$; and hence $u_n \notin N_i$ for all n prime to the characteristic of V_i/N_i. Therefore $u_t \notin N_i$ in this case. Suppose that $q_i > 1$; then $1-a$ is a unit of V_i. Since $(1-a^n) = (1-a)u_n$, we thus have $u_n \in N_i$ if and only if $a^n \equiv 1 \pmod{N_i}$ if and only if n is a multiple of q_i. Therefore $u_t \notin N_i$ in this case either. If $a^n \not\equiv 1 \pmod{N_i}$ for any positive integer n, then $u_n = (1-a^n)/(1-a)$ is a unit and hence is not in N_i for any $n \geq 1$. This proves our assertion.

Now we will prove statement (1). Clearly $R_{M_i} \subseteq V_i$. On the other hand let $a \in V_i$. Then we have seen that there exists an integer $t \geq 1$ such that $1/u_t$ and $a/u_t = b$ are both in R. Since $a \in V_i$, we have $u_t \in V_i$. Therefore, u_t is a unit in V_i, and hence $1/u_t \notin M_i$. Thus $u_t \in R_{M_i}$ and $a = u_t b$ is in R_{M_i}. Thus we have $V = R_{M_i}$.

We now prove statement (2). Let I be an ideal of R, and suppose that I is not contained in any of the M_i. Then for every $i = 1, \ldots, m$, I contains an element c_i such that $c_i \notin M_i$. Since $V_i \not\subseteq V_j$ for $i \neq j$, and since $V_i = R_{M_i}$, it follows that $M_j \not\subseteq M_i$ for $i \neq j$. But the M_i are prime ideals of R. Hence for every i there exists an element $d_i \in \bigcap_{j \neq i} M_j$ such that $d_i \notin M_i$. Let $c = \sum_{i=1}^{m} c_i d_i$; then c is an element of I. Clearly $c \notin M_i$ for every i, and thus c is a unit in every V_i. Thus c is a unit in R and we have $I = R$. This implies that any maximal ideal of R is one of the M_i. Since $M_j \not\subseteq M_i$ for $i \neq j$, it follows that every M_i is a maximal ideal of R.

DEFINITION. An integral domain R is said to be a Prüfer ring if every finitely generated ideal of R is a projective R-module. It is not hard to see that R is a Prüfer ring if and only if R_M is a

valuation ring for every maximal ideal M of R. Thus by Theorem 52,
a finite intersection of valuation rings is a Prüfer ring. Such a finite
intersection is a quasi-semi-local ring by Theorem 52, and for quasi-
semi-local rings finitely generated projectives are free. Thus if R is
a finite intersection of valuation rings, every finitely generated ideal of
R is principal.

The next theorem is a corollary of Theorem 42. It is also a
lemma for Theorem 72.

THEOREM 53. Let R be a completely reflexive ring. Then
the integral closure of R in its quotient field Q is a maximal valuation ring. Furthermore, if V is any valuation ring in Q containing R,
then V is also a maximal valuation ring.

Proof. Suppose that the integral closure of R is not a valuation ring. Then the valuation rings in Q that contain R are not
linearly ordered. Hence there exist two valuation rings V_1 and V_2
in Q that contain R and such that $V_i \not\subset V_j$ for $i \neq j$. By Theorem 52,
$V_1 \cap V_2$ has two distinct maximal ideals, and hence is not a quasi-
local ring. But every ring in Q that contains R is quasi-local by
Theorem 48. This contradiction establishes the fact that the integral
closure of R is a valuation ring.

Let V be a valuation ring in Q that contains R. By Theorem
48, V is a complete valuation ring. Now every finitely generated
ideal of V is principal. This implies that every ideal of V is a flat
V-module. Thus V^{-1} is flat, and hence by Theorem 48, Q/V is an
injective V-module. It follows from Theorem 51 that V is a maximal
valuation ring.

The following theorem is an important lemma for Theorem 83.

THEOREM 54.   Let R be the intersection of two maximal valuation rings in Q. Then K is injective if and only if R is an h-local ring.

Proof. We have $R = V_1 \cap V_2$, where V_1 and V_2 are maximal valuation rings in Q. If one of the V_i is contained in the other, then R is a maximal valuation ring, and K is injective by Theorem 51. Hence we can assume that the V_i are independent valuation rings. Thus by Theorem 52, R has exactly two maximal ideals M_1 and M_2 and $R_{M_i} = V_i$ for $i = 1, 2$. By Theorem 51, K_{M_i} is an injective R_{M_i}-module for $i = 1, 2$. Suppose that R is an h-local ring. Then by Theorem 24 we have inj. $\dim_R K = \sup_{M_i}$ inj. $\dim._{R_{M_i}} K_{M_i} = 0$. Thus K is an injective R-module.

Conversely, assume that K is an injective R-module. Suppose that R is not an h-local ring. Then there exists a non-zero prime ideal P of R such that $P \subset M_1 \cap M_2$. It follows that $R_{M_i} \subset R_P$, and thus R_P is a valuation ring with maximal ideal PR_P, and $PR_P \subset R_{M_i}$. Similarly, we have $PR_P \subset R_{M_2}$; and thus $PR_P \subset R$. Hence $PR_P = P$.

Let S be an R-submodule of Q, $(S \neq Q)$, that contains R. Then we can assume that $S_{M_1} \neq Q$. Hence there exists an element $r \in R$, $r \neq 0$, such that $rS_{M_1} \subset R_{M_1} \subset R_P$. Therefore $(Pr)S_{M_1} \subset PR_P \subset R$, and we have $S_{M_1}^{-1} \neq 0$. Since $S \subset S_{M_1}$, we also have $S^{-1} \neq 0$. This proves that K is an indecomposable R-module. Thus K is an indecomposable injective R-module.

Let $a \neq 0 \in R$ and $x = (1/a + R) \in K$. Then $O(x) = Ra$, and hence by Theorem 38, Ra is an irreducible ideal of R. Let b be any other non-zero element of R. Then we have an exact sequence:

$$0 \to Ra \cap Rb \to Ra \oplus Rb \to Ra + Rb \to 0 \ .$$

Since R is a Prufer ring, $Ra + Rb$ is a projective ideal of R, and thus the exact sequence splits. Therefore $Ra \cap Rb$ is a projective ideal of R. Since R is a quasi-semilocal ring, projective ideals of R are principal ideals of R. Thus $Ra \cap Rb$ is a principal ideal or R. But we have proved that principal ideals are irreducible. Thus we have either $a \in Rb$ or $b \in Ra$, which shows that R is a valuation ring. But R is not a quasi-local ring. This contradiction proves that R is an h-local ring.

9

THE TWO GENERATOR PROBLEM FOR IDEALS

As we have already seen, ideals that can be generated by two
elements are a recurrent theme in the problems we have been considering. This notion takes on a sharper focus in this chapter. The chapter
begins with the description in Theorem 55 of a well-known property of
direct sum decompositions of free modules over quasi-local rings. We
need this theorem as a lemma for Theorem 56. This latter theorem,
(due to H. Bass), introduces the role that ideals generated by two elements will play in the theory of D-rings. The chapter concludes with
Theorem 57 in which we characterize the domains all of whose ideals
can be generated by two elements. It is an open question whether or not
Statement (4) of the theorem is equivalent to the statement that the
domain is Noetherian and satisfies property FD.

THEOREM 55 (Projective Covers). Let R be a com-
mutative, quasi-local ring, F a finitely generated free R-module, and
B a submodule of F. Assume that $F/B = C_1 \oplus \ldots \oplus C_n$, where the
C_i are R-modules. Then there exists a direct sum decomposition:
$F = F_1 \oplus \ldots \oplus F_n$ such that if $B_i = F_i \cap B$, then $B = B_1 \oplus \ldots \oplus B_n$
and $C_i \cong F_i/B_i$.

Proof. Let M be the maximal ideal of R. Let G_i be a free
R-module mapping onto C_i, and assume that G_i has the same number

of generators as a minimal generating set for C_i. If P_i is the kernel

of this map, we have by minimality that $P_i \subset MG_i$. Thus we have a

commutative diagram with exact rows:

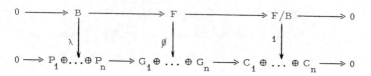

The mapping $\emptyset$ exists because F is a free R-module; and λ is the

restriction of $\emptyset$ to B. Let $P = P_1 \oplus \ldots \oplus P_n$ and $G = G_1 \oplus \ldots \oplus G_n$;

then $P + \emptyset(F) = G$. Since $P \subset MG$, we have $M(G/\emptyset(F)) = \dfrac{MG + \emptyset(F)}{\emptyset(F)}$

$= G/\emptyset(F)$. Thus by the Nakayama Lemma, $G = \emptyset(F)$, and $\emptyset$ is onto.

Therefore λ is onto also.

Since G is free and $\emptyset$ is onto, there exists an R-homomorph-

ism $\theta: G \to F$ such that $\emptyset\theta = 1$ on G. Thus $F = \text{Ker } \emptyset \oplus \theta(G)$ and

$\theta(G) \cong G$. Since $\text{Ker } \emptyset = \text{Ker } \lambda$ and $\theta(P) \subset B$, we have

$B = \text{Ker } \emptyset \oplus \theta(P)$. Let $F_1 = \text{Ker } \emptyset \oplus \theta(G_1)$, $F_2 = \theta(G_2), \ldots, F_n = \theta(G_n)$;

similarly let $B_1 = \text{Ker } \emptyset \oplus \theta(P_1)$, $B_2 = \theta(P_2), \ldots, B_n = \theta(P_n)$. Then

$B_i = F_i \cap B$, and the conclusions of the theorem follow.

DEFINITION. R is said to have <u>property FD</u> if every finitely

generated torsion-free R-module is a direct sum of modules of rank

one.

The next theorem is due to H. Bass [2].

THEOREM 56. Let R be a quasi-local integral domains with

property FD. Then every finitely generated torsion-free R-module of

rank 1 can be generated by two elements.

Proof. Let I be a nonzero, finitely generated ideal of R. It is sufficient to prove that I can be generated by two elements. Let $a_1, \ldots, a_n$ be a minimal generating set for I, and we can assume that $n \geq 2$. Let F be a direct sum of n copies of R; then $x = (a_1, \ldots, a_n) \in F$. Let B be the pure submodule of F generated by x, and thus B is torsion-free of rank 1, and F/B is finitely generated and torsion-free of rank n-1. By property FD we have $F/B = C_1 \oplus \ldots \oplus C_{n-1}$, where C_i is torsion-free of rank 1.

By Theorem 55 there is a direct sum decomposition $F = F_1 \oplus \ldots \oplus F_{n-1}$ such that if $B_i = F_i \cap B$, then $B = B_1 \oplus \ldots \oplus B_{n-1}$ and $C_i \cong F_i/B_i$. However, B is torsion-free of rank 1 and hence is indecomposable. Therefore, we can assume that $B \subset F_1$. Now a change of basis of F comes from a non-singular linear transformation of F. From this it follows that the coordinates of x relative to any basis of F are a generating set for I. Hence, due to the fact that n is the minimal number of elements needed to generate I, B can not be contained in any proper direct summand of F. Thus $F = F_1$, n-1 =1, and thus $n = 2$.

DEFINITION. A Gorenstein ring is a commutative Noetherian ring R such that inj. dim. $_R R$ is finite. If inj. dim. $_R R < \infty$, then inj. dim. $_R R \leq$ the Krull dimension of R. Thus a Noetherian ring is reflexive if and only if it is a Gorenstein ring of Krull dimension 1 by Theorem 40.

THEOREM 57. If R is an integral domain, then the following statements are equivalent:

(1) Every ideal of R can be generated by two elements.

(2) If S is a ring extension of R in Q that is finitely gene-rated as an R-module, then S is a reflexive ring and $\bigcap_n I^n = 0$, for every ideal I of S.

(3) R is a Gorenstein ring of Krull dimension 1; and if M is a maximal ideal of R, then either M is projective, or M^{-1} is a Gorenstein ring of Krull dimension 1.

(4) R is a Noetherian ring such that R_M has property FD for every maximal ideal M of R.

Proof. (1) $\Rightarrow$ (2). Since every finitely generated ring exten-sion R in Q satisfies condition (1), it will be sufficient to prove that R is a reflexive ring and that $\bigcap_n I^n = 0$ for every ideal I of R. It is, of course, trivial that R is a Noetherian domain, and thus $\bigcap I^n = 0$. By Theorem 40 to prove that R is a reflexive ring it will be sufficient to prove that R has Krull dimension 1. Without loss of generality, we can assume that R is a Noetherian local domain with maximal ideal M. We will assume that rank M is greater than 1 and arrive at a contra-diction.

By assumption M can be generated by two elements x and y. Let $\overline{R} = R/Rx$ and $\overline{M} = M/Rx$; then $\overline{R}$ is a Noetherian local ring and its maximal ideal $\overline{M}$ is principal. Since $\bigcap_n \overline{M}^n = 0$, either $\overline{R}$ is an integral domain, or there exists an integer $n > 0$ such that $\overline{M}^n = 0$. In the latter case $M^n \subset Rx$, and by the Krull principal ideal theorem we have rank $M \leq 1$. Therefore, $\overline{R}$ is an integral domain; i.e., Rx is a prime ideal of R.

Now M is not a principal ideal of R since rank $M > 1$, and thus x and y are linearly independent modulo M^2. Because M^2 can be generated by two elements, one of the elements x^2, xy, and y^2 is a linear combination of the other two. If x^2 is a linear combination of xy and y^2 then $M^2 \subseteq Ry$; and hence by the Krull principal ideal theorem rank $M \leq 1$. Thus x^2 is not a linear combination of the other two elements, and by the same reasoning neither is y^2. Therefore, we have $xy = ax^2 + by^2$, where $a, b \in M$.

Since Rx is a prime ideal and $Ry \not\subseteq Rx$ it follows that $b \in Rx$. Therefore, $b = cx$, where $c \in R$; and by cancellation we have $y = ax + cy^2$. But then $y \in M^2$, and this is a contradiction. This proves that R has Krull dimension 1.

$(2) \Rightarrow (3)$. It is sufficient to prove that R is a Noetherian ring. For by Theorem 40, R is then a Gorenstein ring of Krull dimension 1; and if M is not a projective ideal of R, then M^{-1} is a ring that is finitely generated over R, and hence by assumption a Noetherian reflexive ring; i.e., a Gorenstein ring of Krull dimension 1.

Now a reflexive ring is an h-local ring by Theorem 30. Hence by Theorem 26, to prove that R is Noetherian, it is sufficient to prove that R_M is Noetherian for every maximal ideal M of R. The proof now falls naturally into two steps. We prove first that R_M satisfies the condition of (2). We can then assume without loss of generality that R is a quasi-local ring, and prove that R is a Noetherian ring.

Let A be a ring extension of R_M in Q that is finitely generated as an R_M-module. Then there exist elements $x_1, \ldots, x_n$ in A such that $A = R_M[x_1, \ldots, x_n]$. Each x_i is integral over R_M, and hence there exists an element $s \in R - M$ such that $a_i = sx_i$ is integral over R. Let $S = R[a_1, \ldots, a_n]$; then S is finitely generated as an R-module and $S_M = A$.

Since S is finitely generated over R there is a one-to-one correspondence between the maximal ideals of S that do not meet R - M and the maximal ideals of A. Hence if L is a maximal ideal of A, there is a maximal ideal N of S such that $N_M = L$ and $L \cap S = N$. For each positive integer k we have $L^k = (N_M)^k = (N^k)_M$. Since N^k is a primary ideal for the maximal ideal N, we have $\bigcap_k (N^k)_M = (\bigcap_k N^k)_M$. Now $\bigcap_k N^k = 0$, by assumption on S, and therefore, we have $\bigcap_k L^k = \bigcap_k (N^k)_M = (\bigcap_k N^k)_M = 0$. It follows that $\bigcap_k I^k = 0$ for every ideal I of A.

Since S is a reflexive ring by assumption, S is an h-local ring by Theorem 30. But then A is an h-local ring because of the one-to-one correspondence between the prime (respectively maximal) ideals of A and the prime (respectively maximal) ideals of S that do not meet R - M. It is easily verified that $A_L = (S_M)_L = S_N$. But S_N is a reflexive ring by Theorem 32. Hence by the same theorem A is a reflexive ring. Thus we have verified that R_M satisfies the conditions of (2). We thus can assume without loss of generality that R is a quasi-local ring with maximal ideal M.

If M is a projective ideal of R, then it is a principal ideal of R. But then $\bigcap_R M^k = 0$ would imply that R is a discrete valuation ring, and hence certainly a Noetherian ring. Thus we can assume that M is not a projective ideal of R. But then $M^{-1}M = M$, which shows that M^{-1} is a ring R_1 that is an extension of R in Q. By Theorem 35,

$R_1/R \cong R/M$ and R_1 can be generated by two elements over R. Hence by assumption R_1 is a reflexive ring.

If J is an ideal of R that is not a principal ideal of R, then J is actually an ideal of R_1. For we have $JJ^{-1} \subset M$, and therefore $(R_1 J)J^{-1} \subset R_1 M = M$. Thus $(R_1 J) \subset J^{-1-1} = J$.

We proceed to establish several properties of R and R_1.

(a) <u>M is a principal ideal of R_1.</u>

If I is an ideal of R_1, we define $I^{\#} = \{x \in Q \mid xI \subset R_1\}$; thus $I^{\#}$ is the dual of I with respect to R_1. Suppose M is not a principal ideal of R_1. Since R_1 is a finitely generated module over the quasi-local ring R, R_1 is a quasi-semi-local ring, and finitely generated projective R_1-modules are free. Thus M is not a projective ideal of R_1, and we have $MM^{\#} \neq R_1$. Thus there exists a maximal ideal P of R_1 such that $MM^{\#} \subset P$. Hence $(P^{\#}M)M^{\#} \subset P^{\#}P \subset R_1$, and therefore $P^{\#}M \subset M^{\#\#}$. Since R_1 is a reflexive ring, we have $M^{\#\#} = M$, and hence $P^{\#}M \subset M$. Thus $P^{\#} \subset M^{-1} = R_1$, and it follows that $P^{\#} = R_1$. But this contradicts $P^{\#\#} = P$, and hence M is a principal principal ideal of R_1.

We have just shown that there exists an element $a \in M$ such that $M = R_1 a$. Thus we have the relations:

$$M^2 = Ma \subsetneq Ra \subsetneq M.$$

It also follows that if $x \in M$, then $x/a \in R_1$.

(b) If R_1 is not a quasi-local ring with maximal ideal $N \underset{\neq}{\supset} M$, then R_1 is a principal ideal domain.

Since $R_1/R \cong R/M$, every maximal ideal of R_1 contains M, and there are at most two such ideals. If R_1 is quasi-local with maximal ideal M, then, since M is a principal ideal of R_1 and $\bigcap_k M^k = 0$, it follows that R_1 is a discrete valuation ring. Thus we may suppose that R_1 has two maximal ideals N_1 and N_2. Then $M = N_1 \cap N_2 = N_1 N_2$. It follows that N_1 and N_2 are projective ideals and thus principal ideals of R_1. Since $\bigcap_k N_1^k = 0$ and $\bigcap_k N_2^k = 0$, we see that N_1 and N_2 are the only nonzero prime ideals of R_1. Hence every prime ideal of R_1 is principal, and thus by Theorem 49, R_1 is a Noetherian ring. But then R_1 is a principal ideal domain which establishes our assertion.

If R_1 is a principal ideal ring, then every non-principal ideal of R, being an ideal of R_1, is isomorphic to R_1 and hence can be generated by two elements over R. Thus R is a Noetherian ring. Therefore, we can assume that R_1 is not a principal ideal ring. We then have by (b) that R_1 is a quasi-local ring with maximal ideal $N \underset{\neq}{\supset} M$. If N is a principal ideal of R_1, then R_1 is a discrete valuation ring. Thus we can assume that N is not a principal ideal of R_1. Let $R_2 = N^{\#}$, the dual of N with respect to R_1. Then R_2 is a ring that is generated by two elements over R_1, since R_1 is a reflexive ring.

(c) $N = R_2 a$, where $M = R_1 a$.

We have $M \subsetneq N \subsetneq R_1$; hence by taking duals with respect to R we have $M \subsetneq N^{-1} \subsetneq R_1$. It is easy to see that N^{-1} is an R_1-ideal, and hence $N^{-1} \subset N$. Thus we have $M \subsetneq N^{-1} \subset N \subsetneq R_1$. Since R_1/M has dimension 2 over R/M, we must have $N^{-1} = N$. It is easy to see that $N^{\#} = (MN)^{-1}$. Therefore, $R_2 = N^{\#} = (MN)^{-1} = (aN)^{-1} = N^{-1}a^{-1} = Na^{-1}$, and hence $N = R_2 a$.

The preceding remarks concerning (b) and (c) have established the following statement:

(d) <u>There exists a chain of quasi-local rings</u> $R \subset R_1 \subset R_2 \subset \ldots \subset R_n \subset \ldots$ <u>such that each</u> R_i <u>is a reflexive ring</u> <u>in</u> Q; R_{i+1} <u>is generated by two elements over</u> R_i <u>and hence is</u> <u>finitely generated over</u> R; <u>if</u> N_i <u>is the maximal ideal of</u> R_i, <u>then</u> R_{i+1} <u>is the dual of</u> N_i <u>with respect to</u> R_i; $N_i = R_{i+1}a$; <u>and</u> $M \subset N \subset N_2 \subset \ldots$. <u>The chain of rings</u> R_n <u>terminates if and only if</u> R_n <u>is a principal ideal ring for some</u> n.

(e) <u>R is a Noetherian ring</u>

Suppose that R is not a Noetherian ring. Then by Theorem 49, R has a non-finitely generated prime ideal P. The ideal P is actually an R_n-ideal for each integer n. If R_n were a principal ideal ring, then P would be isomorphic to R_n, and hence finitely generated over R. This contradiction shows that the chain of rings in (d) does not terminate.

Since each R_i is integral over R, there exists a chain $P \subset P_1 \subset P_2 \subset \ldots \subset P_n \subset \ldots$, where each P_i is a prime ideal of R_i,

$P_{i+1} \cap R_i = P_i$, and $P_i \cap R = P$. It follows that $P_i \neq N_i$ for each i, and hence no power of a lies in any P_i. Let b be a nonzero element of P. Since $b \in M$, b/a is an element of P_1. We have $(b/a)a = b \in P_1$, and hence $b/a \in P_1$. Therefore, $(b/a^2) \in R_2$ and $(b/a^2)a^2 \in P_2$, which shows that $b/a^2 \in P_2$. By an obvious induction, we have $b/a^n \in R_n$ for every n. But since P is an R_n-ideal, it follows that $(b/a^n)b = b^2/a^n \in P$ for every integer n. Hence $b^2 \in \bigcap_n Pa^n \subset \bigcap_k M^k$ = 0. This contradiction shows that R is a Noetherian ring.

$(3) \Rightarrow (1)$. Since R is a Noetherian ring of Krull dimension 1, every nonzero ideal of R is contained in only a finite number of maximal ideals. Thus by Theorem 26, it is sufficient to prove that every ideal of R_M can be generated by two elements for every maximal ideal M of R. Since R_M satisfies the conditions of (3) by Theorem 31, we can assume without loss of generality that R is a Noetherian local ring with maximal ideal M. We can assume that M is not a principal ideal of R.

Let I be a nonzero ideal of R. Since I is an M-primary ideal, R/I has finite length which we will denote by $\ell(R/I)$. The proof of (a) in $(2) \Rightarrow (3)$ only used the assumption that R and M^{-1} were reflexive rings. Thus $M = R_1 a$ is isomorphic to R_1, and hence can be generated by two elements. Hence $\ell(M/M^2) = 2$ and thus $\ell(R/M^2) = 3$. As we have seen in (a) we have $M^2 \subsetneq Ra \subsetneq M \subsetneq R$. Hence $\ell(R/Ra) = 2$.

Assume that I is a nonprincipal ideal of R. From the exact sequence

$$0 \to \frac{Ra}{MI} \to \frac{R}{MI} \to \frac{R}{Ra} \to 0$$

we conclude that $\ell(R/MI) = \ell(R/Ra) + \ell(Ra/MI)$. But $MI = R_1 aI = aI$, and hence $Ra/MI = Ra/Ia \cong R/I$. Thus we have $\ell(R/MI) = 2 + \ell(R/I)$. From the exact sequence

$$0 \to \frac{I}{MI} \to \frac{R}{MI} \to \frac{R}{I} \to 0$$

we conclude that $\ell(R/MI) = \ell(I/MI) + \ell(R/I)$. Comparing the two equations for $\ell(R/MI)$ we see that $\ell(I/MI) = 2$. Thus I can be generated by two elements.

$(2) \Rightarrow (4)$. Since we have already proved that $(1), (2),$ and (3) are equivalent we can assume that R is a Noetherian local domain of Krull dimension 1 such that every ideal of R can be generated by two elements. Let A be a finitely generated, torsion-free R-module of rank greater than 1. We shall assume that A is indecomposable and obtain a contradiction.

If $f \in A' = \operatorname{Hom}_R(A, R)$, then $f(A)$ is an ideal of R. The trace ideal I of A is defined to be the sum of all of the $f(A)$ for $f \in A'$.

Case I: $I = R$.

In this case there exists elements $x_1, \ldots, x_n \in A$ and $f_1, \ldots, f_n \in A'$ such that $\sum_{i=1}^{n} f_i(x_i) = 1$. Since R is a local ring, there is an integer i, $1 \leq i \leq n$ such that $f_i(x_i)$ is a unit of R. Hence there are elements $x \in A$ and $f \in A'$ such that $f(x) = 1$. If we define $g: R \to A$ by $g(r) = rx$, then fg is equal to the identity on R. Hence Rx is a direct summand of A. This is a contradiction, and thus $I \neq R$.

Case II: $I \subset M$, the maximal ideal of R.

If M is a principal ideal of R, then R is a discrete valuation ring. But then A is a free R-module of rank > 1, and hence A can not be indecomposable. Thus M is not a projective ideal of R, and $R_1 = M^{-1}$ is a ring. We will show that A is an R_1-module.

Let $q \in R_1$, $x \in A$, and $f \in A'$. We define $(qx)(f) = qf(x)$. Since $M^{-1} \subset I^{-1}$ and $f(x) \in I$, it follows that $qf(x) \in R$. Thus $qx \in A''$. But R is a reflexive ring and so $A \cong A''$. Thus $qx \in A$, and we have defined an operation of R_1 on A extending that of R; in other words, A is an R_1-module.

Using the chain of rings $R \subset R_1 \subset R_2 \subset \ldots \subset R_n \subset \ldots$ established in part (d) of $(2) \Rightarrow (3)$, we repeat our procedure and see that A is an R_n-module for each n. If the chain terminates at R_n, then R_n is a principal ideal domain, and A is a free R_n-module, necessarily decomposable. Thus the chain does not terminate at any n.

Let $S = \bigcup_{n=0}^{\infty} R_n$; then S is a ring and S is not finitely generated as an R-module. Furthermore, A is an S-module by the preceding remarks. Let $x \in A$, $x \neq 0$; then $Sx \subset A$, and Sx is an R-submodule of A isomorphic to S. But R is Noetherian, and hence every R-submodule of A is finitely generated. This contradiction shows that A is not indecomposable, and (4) is established.

$(4) \Rightarrow (1)$. Let M be a maximal ideal of R. By Theorem 56 every ideal of R_M can be generated by two elements. As in the proof of $(1) \Rightarrow (2)$ we see that R_M has Krull dimension 1. Hence R has Krull dimension 1, and thus is an h-local ring. We can now apply

Theorem 26 to conclude that every ideal of R can be generated by two elements.

10

NOETHERIAN D-RINGS

We begin this chapter by proving some elementary properties of D-rings. Theorem 58 is a lemma for Theorem 59, and the latter establishes the hereditary nature of property D. Theorem 60 is a lemma for Theorem 61, and Theorem 61 states that a quasi-local D-ring is complete in the R-topology. Theorem 62 is an old theorem due to F. K. Schmidt [34]. We use this theorem to prove Theorem 63 which settles a question raised by Kaplansky [14]. In its turn Theorem 63 helps to prove Theorem 64 which states that the only Noetherian D-rings are the rings of type II. Schmidt's theorem provides the explanation of why Noetherian rings of type I do not exist.

DEFINITION. An integral domain R is said to have property D, or to be a D-ring, if every torsion-free R-module of finite rank is a direct sum of modules of rank 1.

THEOREM 58. Let S be an extension ring of R contained in the quotient field Q of R, and let A be a torsion-free S-module. If $A = A_1 \oplus A_2$, where the A_i are R-modules, then the A_i are also S-modules.

Proof. Choose $x \in A_1$, and $s \in S$, $s \neq 0$; then $s = a/b$ where $a, b \in R$ and $b \neq 0$. Since $sx \in A$, we have $sx = x_1 + x_2$, where

$x_i \in A_i$ for $i = 1, 2$. Then $ax = b(sx) = bx_1 + bx_2$. Hence $bx_2 \in A_1 \cap A_2 = 0$, and thus $x_2 = 0$. Therefore $sx \in A_1$, and A_1 is an S-module. Similarly, A_2 is an S-module.

THEOREM 59. Let R be a D-ring and S a subring of Q that contains R. Then S is also a D-ring.

Proof. This is an immediate consequence of Theorem 58.

THEOREM 60. Let R be a quasi-local ring and

$$0 \to R \to A \to Q \to 0$$

an exact sequence of R-modules. Then A is a decomposable R-module if and only if R is a direct summand of A.

Proof. A is a torsion-free R-module of rank 2. Suppose that $A = A_1 \oplus A_2$, where A_i is a torsion-free R-module of rank 1 for $i = 1, 2$. Applying the functor $K \otimes_R \cdot$ to the given exact sequence we obtain $K \cong (K \otimes_R A_1) \oplus (K \otimes_R A_2)$. But K is indecomposable by Theorem 18, and hence we can assume that $K \otimes_R A_2 = 0$. Therefore, by exact sequence (II), it follows that A_2 is isomorphic to Q. Thus A_2 maps isomorphically onto Q in the given exact sequence. Consequently, $A = R + A_2$ and $R \cap A_2 = 0$, and we see that R is a direct summand of A.

THEOREM 61. Let R be a quasi-local ring such that every torsion-free R-module of rank 2 that is _not_ finitely generated is a direct sum of modules of rank 1. Then R is complete in the R-topology.

Proof. An extension of R by Q is not finitely generated and hence splits by Theorem 60. Thus $\text{Ext}_R^1(Q, R) = 0$, and thus R is complete by Theorem 9.

The following theorem is due originally to F. K. Schmidt [34]. However, the elementary proof given here was communicated to me by Barbara Osofsky.

THEOREM 62. Let R be a principal ideal domain with more than 1 nonzero prime ideal. Then R_P is not a complete ring for any nonzero prime ideal P of R.

Proof. Let P_1 and P_2 be two distinct nonzero prime ideals of R and suppose that R_{P_1} is a complete ring. By localizing with respect to the complement of $P_1 \cup P_2$, we can assume that P_1 and P_2 are the only nonzero prime ideals of R. Now P_1 and P_2 are generated by prime elements p_1 and p_2, and $p_1 + p_2$ can not be a product of powers of p_1 and p_2. Thus $p_1 + p_2$ is a unit in R, and without loss of generality we can assume that $p_1 + p_2 = 1$.

Case I: The characteristic of $R/(p_1)$ is not equal to 2.

Let X be an indeterminate. Then $X^2 - p_2$ is congruent to $X^2 - 1$ modulo (p_1), and hence factors into $(X-1)(X+1)$ over $R/(p_1)$. These factors are relative prime over $R/(p_1)$ since the characteristic is $\neq 2$. But R_{P_1} is complete, and thus by Hensel's Lemma, $X^2 - p_2$ factors over the ring R_{P_1}. It follows that p_2 is a square in R_{P_1} and hence a square in R. This contradicts the fact that p_2 is a prime.

Case II. The characteristic of $R/(p_1)$ is 2.

Now $X^3 - p_2$ is congruent to $X^3 - 1$ modulo (p_1), and hence factors into $(X-1)(X^2 + X + 1)$ over $R/(p_1)$. These factors are relatively prime over $R/(p_1)$, since the characteristic of $R/(p_1)$ is 2. Since R_{P_1} is complete, it follows from Hensel's Lemma that $X^3 - p_2$

factors over R_{P_1}. But this implies that p_2 is a cube in R_{P_1} and hence a cube in R. This contradicts the fact that p_2 is prime.

Thus we have a contradiction in both cases, and this proves that R_{P_1} can not be a complete ring.

THEOREM 63. The following statements are equivalent:

(1) R is a complete Noetherian local domain of Krull dimension 1.

(2) R is a Noetherian domain, and every torsion-free R-module of rank **2** which is not finitely generated is a direct sum of modules of rank 1.

Proof. (1) $\Rightarrow$ (2). This is an immediate consequence of Theorem 44.

(2) $\Rightarrow$ (1). We will assume that R is not a Noetherian, local domain of Krull dimension 1 and arrive at a contradiction. By the Krull principal ideal theorem every nonzero element of R that is not a unit is contained in a rank 1 prime ideal of R. Thus R has at least two distinct rank 1 prime ideals P_1 and P_2. By Theorem 58 every Noetherian ring between R and Q has property (2). Thus we can localize with respect to the complement of $P_1 \cup P_2$, and hence we can assume that P_1 and P_2 are the only nonzero prime ideals of R. The integral closure of R in Q is then a principal ideal domain with more then 1 nonzero prime ideal, and hence we can assume that R is a principal ideal domain. But R_{P_1} is complete by Theorem 61, and this fact contradicts Theorem 62. Therefore, R is a Noetherian, local domain of Krull dimension 1. R is complete by Theorem 61.

DEFINITION. We say that an integral domain R is a <u>ring of type II</u>, if R is a complete, Noetherian, local domain such that every ideal of R can be generated by two elements. By Theorem 57 a ring of type II has Krull dimension 1.

The next theorem completely characterizes Noetherian D-rings.

THEOREM 64. The following statements are equivalent:

(1) R is a Noetherian D-ring.

(2) R is a ring of type II.

If either of these conditions is true, then R is a completely reflexive ring.

Proof. (1) $\Rightarrow$ (2) By Theorem 63, R is a complete Noetherian local domain of Krull dimension 1. By Theorem 57, every ideal of R can be generated by two elements, and R is a completely reflexive ring.

(2) $\Rightarrow$ (1). By Theorem 44, every torsion-free R-module of finite rank is a direct sum of a finitely generated R-module and a divisible R-module. The divisible module is isomorphic to a direct sum of copies of Q; and by Theorem 57, the finitely generated module is isomorphic to a direct sum of modules of rank 1. Thus R has property D.

11

QUASI-LOCAL D-RINGS

Kaplansky initiated the study of D-rings when be proved that a maximal valuation ring has property D [12]. Theorem 65 gives a new proof of this result and provides a converse when R is a valuation ring. Theorems 66 through 69 state some properties of rings in which every finitely generated ideal can be generated by two elements. Theorems 70 and 71 explore the relationships between quasi-local D-rings and completely reflexive rings. Theorem 72, the most important theorem of the chapter, proves that the integral closure of a quasi-local D-ring is a maximal valuation ring. Theorem 75 is a partial converse of Theorem 72 and provides a method for constructing quasi-local D-rings that are not maximal valuation rings. The chapter closes with some examples.

THEOREM 65. Let R be a valuation ring. Then R is a D-ring if and only if R is a maximal valuation ring.

Proof. If R is a maximal valuation ring, then R has property D as an immediate consequence of Theorem 51 (2) which states that a pure submodule of a torsion-free R-module of finite rank is a direct summand.

Conversely, suppose that R is a valuation ring with property D. Then R is complete in the R-topology by Theorem 61.

Let I and J be nonzero ideals of R, and consider an exact sequence of the form:

$$0 \to I \to A \xrightarrow{p} J \to 0 ,$$

where A is an R-module. Since A is a torsion-free R-module of rank 2, we have by property D that $A = A_1 \oplus A_2$, where A_1 and A_2 are torsion-free R-modules of rank 1. Since R is a valuation ring, the R-submodules of J are linearly ordered. Thus we can assume that $p(A_1) \subseteq p(A_2)$. But $J = p(A_1) + p(A_2)$, and so $J = p(A_2)$. It follows that $A = I + A_2$. Since an R-homomorphism between two torsion-free R-modules of rank 1 is either 0 or a monomorphism, we have $A_2 \cap \text{Ker } p = 0$; that is $A_2 \cap I = 0$. Thus $A = I \oplus A_2$, and the given exact sequence splits. This shows that $\text{Ext}_R^1(J, I) = 0$.

But $\text{Ext}_R^1(J, I) \cong \text{Ext}_R^2(R/J, I)$. Hence we have $\text{inj.dim}_R I = 1$ for every nonzero ideal I of R. Thus R is a maximal valuation ring by Theorem 51 (3).

THEOREM 66. Let R be a quasi-local ring with maximal ideal M such that every finitely generated ideal of R can be generated by two elements. Then the following statements are true.

(1) If A is an R-submodule of Q generated by a finite number of elements, then two of these elements already generate A.

(2) If C is an R-submodule of Q, then $\dim_{R/M} C/MC \leq 2$.

(3) M^{-1} can be generated by two elements.

Proof. (1) Let $x_1, \ldots, x_n$ be the generators of A, and $\overline{x}_1, \ldots, \overline{x}_n$ their images in A/MA. Since A is isomorphic to an ideal of R, it can be generated by two elements, and hence $\dim_{R/M} A/MA \leq 2$. Thus we can assume that $\overline{x}_1, \overline{x}_2$ generate A/MA. Let B be the R-module generated by x_1 and x_2. Then A = B + MA, and thus $M(A/B) = (B + MA)/B = A/B$. Therefore, by the Nakayama Lemma we have A = B; that is, A is generated by x_1 and x_2.

(2) Suppose that $\dim_{R/M} C/MC > 2$. Then there exist elements x, y, z in C such that their images $\overline{x}, \overline{y}, \overline{z}$ in C/MC are linearly independent over R/M. By (1) we can assume that z is a linear combination of x and y. But then $\overline{x}, \overline{y}, \overline{z}$ are linearly dependent over R/M. This contradiction shows that $\dim_{R/M} C/MC \leq 2$.

(3) If $MM^{-1} = R$, then M is a projective ideal of R, and hence a principal ideal of R, since R is a quasi-local ring. But then M^{-1} is also principal. Thus we can assume that $MM^{-1} \neq R$; and hence $MM^{-1} = M$. We then have by (2) that $\dim_{R/M} M^{-1}/M \leq 2$. Therefore, there exists an element x in M^{-1} such that the images of 1 and x generate M^{-1}/M. Let B be the R-module generated by 1 and x. Then $M^{-1} = B + M = B$, and so M^{-1} can be generated by two elements.

THEOREM 67. Let R be an integral domain such that every finitely generated ideal of R can be generated by two elements. Then the integral closure of R in Q is a Prüfer ring.

Proof. Clearly every finitely generated torsion-free R-module of rank 1 is isomorphic to an ideal of R, and can thus be generated by two elements. This property is obviously inherited by every ring between R and Q. Thus without loss of generality we can assume that R is an integrally closed, quasi-local ring, and we must prove that R is a valuation ring.

Let M be the maximal ideal of R, and V a valuation ring in Q that dominates R; that is, R $\subset$ V and N $\cap$ R = M, where N is the maximal ideal of V. We will prove that R = V.

Suppose that R $\neq$ V. If every unit of V is contained in R, then V = R. Hence there exists a unit x of V which is not in R. Let A be the R-module generated by 1, x, and x^2. By assumption, A can be generated by two elements. Hence by Theorem 66, two of the elements 1, x, x^2 generate A. Since R is integrally closed, 1 and x can not generate A.

In fact, 1 and x^2 generate A. For if x and x^2 generate A, then 1 = ax + bx^2, where a, b $\in$ R. If both a and b are in M, then 1 $\in$ VM $\subset$ N and this is a contradiction. Hence either a or b is not in M. Since 1 and x do not generate A, we must have b $\in$ M, and thus a is a unit in R. But this shows that 1 and x^2 generate A.

Thus we have shown that x = c + dx^2, where c, d $\in$ R. We must have d $\in$ M, since 1 and x do not generate A. However, c $\notin$ M, since x is a unit in V. But then $\frac{1}{x}$ is integral over R, and hence $\frac{1}{x}$ $\in$ M. Therefore, 1 = x $\cdot$ ($\frac{1}{x}$) $\in$ VM $\subset$ N. This contradiction shows that R = V.

THEOREM 68. Let R be a D-ring. Then the integral closure of R is a Prüfer ring.

Proof. If M is a maximal ideal of R, then R_M has property D by Theorem 59. If the integral closure of R_M is a Prüfer ring for every maximal ideal M of R, then the integral closure of R is a

Prüfer ring. Thus we can assume without loss of generality that R is a quasi-local ring. But then by Theorem 56 every finitely generated ideal of R can be generated by two elements, and hence by Theorem 67, the integral closure of R is a Prüfer ring.

THEOREM 69. Let R be an integral domain such that every finitely generated ideal of R can be generated by two elements. Let x be an element of Q that is integral over R. Then there exist elements $a, b \in R$ such that $x^2 = ax + b$.

Proof. We will suppose that x^2 is not an element of $Rx + R$ and arrive at a contradiction. Let $I = \{r \in R \mid rx^2 \in Rx + R\}$. Then I is contained in a maximal ideal M of R. If $x^2 \in R_M x + R_M$, then there exists an element $s \in R - M$ such that $sx^2 \in Rx + R$. However, since $s \notin I$, we see that $x^2 \notin R_M x + R_M$. Now every finitely generated ideal of R_M can be generated by two elements, and x is integral over R_M. Thus without loss of generality we can assume that R is a quasi-local ring with maximal ideal M.

Let A be the R-module generated by $1, x,$ and x^2. By Theorem 66 two of these elements generate A. Since $x^2 \notin Rx + R$, 1 and x do not generate A. As in the proof of Theorem 67, we have in all cases that 1 and x^2 generate A. Therefore, we have $x \in Mx^2 + R$. An easy induction proves that for any integer $n > 0$, $x, x^2, \ldots, x^{n-1}$ are all elements of $Mx^n + R$. Since x is integral over R, there exists an equation of the form

$$x^n = b_{n-1} x^{n-1} + \ldots + b_1 x + b_0$$

where $b_i \in R$. Thus we have $x^n \in Mx^n + R$. But this implies that

$x^n \in R$. Since $x \in Mx^n + R$, we have $x \in R$. Thus 1 generates A, and this contradiction shows that $x^2 \in Rx + R$.

THEOREM 70. Let R be a quasi-local D-ring. If the maximax ideal M of R is not a principal ideal of R, and if $M^{-1} \neq R$, then R is a completely reflexive ring.

Proof. We will first prove that K is an essential extension of M^{-1}/R. For this purpose it will be sufficient to take an R-module A such that $R \subsetneq A \subset Q$, and show that $M^{-1} \subset A$. We will suppose that $M^{-1} \not\subset A$ and arrive at a contradiction. By Theorems 56 and 66, M^{-1} can be generated by two elements; and hence by Theorem 34, M^{-1}/R is a simple R-module. Thus we have $A \not\subset M^{-1}$, and hence we can choose an element $y \in A$ such that $y \notin M^{-1}$. Choose an element $x \in M^{-1}$ such that 1 and x generate M^{-1}. Since $R \subset A$, we have $x \notin A$.

Let B be the R-module generated by $1, x,$ and y. By Theorem 66, two of these elements generate B. Since $x \notin A$ and $y \notin M^{-1}$, it follows that there exist elements $a, b \in M$ such that $1 = ax + by$. Since M is not a principal ideal of R, we have $M^{-1}M = M$. Thus $ax \in M$ and $by = 1 - ax$ is a unit in R. Hence there exists an element $r \in M$ such that $y = 1/r$. Since $Rr \subset M$, we have $M^{-1} \subset Ry \subset A$. This contradiction shows that K is an essential extension of M^{-1}/R.

Since $M^{-1}/R \cong R/M$, and K is an essential extension of M^{-1}/R we have an embedding $K \subset E(R/M)$, the injective envelope of R/M. We will prove that $K = E(R/M)$. Now every nonzero principal ideal of R is the annihilator of an element of K, and hence of an element of $E(R/M)$. But $E(R/M)$ is an indecomposable injective module, and thus every principal ideal of R is irreducible by Theorem 38.

Let $I \neq 0$ be an ideal of R, and consider an extension C of R by I:

$$0 \rightarrow R \rightarrow C \rightarrow I \rightarrow 0$$

Then C is a torsion-free R-module of rank two; and since R has property D, we have $C = C_1 \oplus C_2$, where C_1, C_2 are torsion-free R-modules of rank 1. Let J_1 and J_2 be the images of C_1 and C_2 in I. Then $J_1 + J_2 = I$ and $J_1 \cap J_2 \cong R$. Hence $J_1 \cap J_2$ is a principal ideal of R. Since we have shown that principal ideals of R are irreducible we can assume that $J_1 \subset J_2$. But then $I = J_2$, which proves that $C = C_2 + R$. Since C_2 has rank 1 and maps onto I, we must have $C_2 \cap R = 0$; that is, $C = C_2 \oplus R$.

Therefore, the preceding exact sequence splits, and this shows that $\text{Ext}_R^1(I, R)$. Since $\text{Ext}_R^2(R/I, R) \cong \text{Ext}_R^1(I, R)$, we have $\text{Ext}_R^2(R/I, R) = 0$. Thus inj. $\dim_R R = 1$. Hence K is an injective R-module, and since $K \subset E(R/M)$, we have $K = E(R/M)$. Thus R is a reflexive ring by Theorem 29.

Now a quasi-local D-ring is complete in the R-topology by Theorem 61. Hence by Theorem 46, R is a completely reflexive ring.

THEOREM 71. Let R be a quasi-local D-ring and let F be the integral closure of R. If $R \neq F$, then there exists a ring S such that $R \subset S \subsetneq F$ and S is a completely reflexive ring.

Proof. Let M be the maximal ideal of R, let $x \in F - R$, and let $S = R + Mx$. By Theorem 56 every finitely generated ideal of R can be generated by two elements, and so by Theorem 69, S is a ring. Let $N = M + Mx$; then N is a proper ideal of S. The elements of S that are not in N are units in F, and hence units in S, since F is the integral closure of S. Thus S is a quasi-local ring with maximal ideal N. By Theorem 59, S has property D.

Let $T = R + Rx$; by Theorem 69, T is a ring. Since $x \notin S$, it follows that $T \neq S$. Let $N^{\#} = \{q \epsilon Q \mid qN \subset S\}$. Since $TN = N$, we see that $T \subset N^{\#}$, and thus $N^{\#} \neq S$. If N is a principal ideal of S, then there exists an element $a \epsilon S$ such that $N = Sa$. But then $Sa = TN = Ta$, and thus $S = T$. This contradiction shows that N is not a principal ideal of S. We have verified that S satisfies the hypotheses of Theorem 70, and thus S is a completely reflexive ring.

THEOREM 72. Let R be a quasi-local D-ring. Then R is a complete ring, and the integral closure F of R is a maximal valuation ring.

Proof. R is complete in the R-topology by Theorem 61. If $R \neq F$, then by Theorem 71 there is a completely reflexive ring S such that $R \subset S \subset F$. But then F is the integral closure of S, and hence by Theorem 53, F is a maximal valuation ring. Thus we can assume that $R = F$; that is, R is integrally closed. By Theorem 68, R is a Prüfer ring. Since quasi-local Prüfer rings are valuation rings, R is a valuation ring. Therefore, R is a maximal valuation ring by Theorem 65.

THEOREM 73. Let R be a D-ring. Let P be a prime ideal of R, and let F be the integral closure of R. Then there is only one prime ideal of F lying over P.

Proof. R_P is a quasi-local D-ring by Theorem 59. The integral closure of R_P is F_P. Thus by Theorem 72, F_P is a maximal valuation ring. From this it follows immediately that F has only one prime ideal lying over P.

The following theorem is due to David Eisenbud. It is a lemma for Theorem 75.

THEOREM 74.   Let $R \subset S$ be commutative rings such that S is a finitely generated R-module. Let C be an R-module such that $\text{Hom}_R(S, C)$ is an injective S-module. Then C is an injective R-module.

Proof.   Let E be the injective envelope of C as an R-module. Since C is a submodule of E, we can consider $\text{Hom}_R(S, C)$ to be an S-submodule of $\text{Hom}_R(S, E)$. We will prove first that $\text{Hom}_R(S, C) = \text{Hom}_R(S, E)$. Since $\text{Hom}_R(S, C)$ is an injective S-module, it is a direct summand of $\text{Hom}_R(S, E)$. Thus to prove the equality of the two modules it is sufficient to prove that $\text{Hom}_R(S, E)$ is an essential extension of $\text{Hom}_R(S, C)$ as an R-module.

Let f be a nonzero element of $\text{Hom}_R(S, E)$. Then f(S) is a nonzero, finitely generated R-submodule of E. Since E is an essential extension of C, there exists an element $r \in R$, $r \neq 0$, such that rf(S) is a non-zero submodule of C. Then rf is a non-zero element of $\text{Hom}_R(S, C)$. Therefore, $\text{Hom}_R(S, E)$ is an essential extension of $\text{Hom}_R(S, C)$, and the two modules are equal.

Suppose now that C is not an injective R-module. Then $C \neq E$, and hence there exists an R-homomorphism $g: R \to E$ such that $g(R) \not\subset C$. Since E is an injective R-module, g can be extended to an R-homomorphism $f: S \to E$; that is, $f \in \text{Hom}_R(S, E)$. But $f \notin \text{Hom}_R(S, C)$ which is a contradiction. Thus C is an injective R-module.

The following theorem is a partial converse of Theorem 72. It provides a way of constructing rings with property D.

THEOREM 75. Let R be a quasi-local integral domain with maximal ideal M, and let F be the integral closure of R. If F is a maximal valuation ring that is generated as an R-module by two ele - ments, and if $MF = M$, then R is a D-ring.

Proof. Since a maximal valuation ring is a D-ring by Theorem 65, we can assume that $R \neq F$. We will first prove that $F = M^{-1}$. Since $MF = M$, we have $F \subset M^{-1}$.

Now M is not a principal ideal of R. For if $M = Ra$, for some $a \in R$, then $Ra = MF = Fa$, and hence $R = F$, which contradicts our assumption. Now invertible ideals in a quasi-local ring are princi-pal ideals, and thus M is not an invertible ideal of R. Therefore, we have $M^{-1}M = M$, and thus M^{-1} is a ring. Furthermore, since $F \subset M^{-1}$, and F is a valuation ring, M^{-1} is also a valuation ring.

Let $m(F)$ and $m(M^{-1})$ be the maximal ideals of F and M^{-1}, respectively. Then we have $m(M^{-1}) \subset m(F)$. If $m(M^{-1}) = m(F)$, then $M^{-1} = F$, and that is our contention. Hence we will assume that $m(M^{-1}) \neq m(F)$, and arrive at a contradiction.

We have $M \subset m(M^{-1}) \subsetneq m(F) \subsetneq F$. Since F can be gene-rated by two elements as an R-module, and since $MF = M$, we have $\dim_{R/M} F/M \leq 2$. Hence the above chain of inequalities shows that $M = m(M^{-1})$. Thus M is a prime ideal of F. Let x be an element of $m(F)$ such that $x \notin M$. Then $x^2 \notin M$; and we have $M \subsetneq Fx^2 \subsetneq Fx \subsetneq F$, since F is a valuation ring. This contradicts $\dim_{R/M} F/M \leq 2$ and proves that $F = M^{-1}$.

We will next prove that R is a completely reflexive ring. Since $M^{-1-1} = M$, we have $F = F^{-1-1}$. Thus F is a reflexive R-module.

But F can be generated by two elements, and thus by Theorem 33 we have $\text{Ext}_R^1(F, R) = 0$. Therefore, if we apply the functor $\text{Hom}_R(F, \cdot)$ to exact sequence (I) we obtain an exact sequence

$$0 \to \text{Hom}_R(F, R) \to \text{Hom}_R(F, Q) \to \text{Hom}_R(F, K) \to 0.$$

Now $\text{Hom}_R(F, R) \cong F^{-1} = M$, and $\text{Hom}_R(F, Q) \cong Q$. Thus $\text{Hom}_R(F, K)$ is isomorphic to Q/M as an F-module. Therefore, by Theorem 51 (3), $\text{Hom}_R(F, K)$ is an injective F-module. It now follows from Theorem 74 that K is an injective R-module.

Since $M^{-1}/R \neq 0$, K contains a copy of the simple R-module R/M. Thus by Theorem 29, R is a reflexive ring.

Now F is complete in the F-topology, and F is isomorphic to an ideal of R. It therefore follows from Theorems 14 and 15 that R is complete in the R-topology. Thus we have by Theorem 46 that R is a completely reflexive ring.

Let A be an indecomposable, torsion-free R-module of finite rank. To prove that R has property D we must show that rank $A = 1$. We can assume that A is a reduced R-module. Since R is a completely reflexive ring, A is a reflexive R-module; that is, $A \cong A''$.

Let I be the trace ideal of A; by definition:
$I = \{ \sum_{i=1}^{n} f_i(x_i) \mid f_i \in A' \text{ and } x_i \in A \}$. If $I = R$, then since R is a quasi-local ring there exist elements $f \in A'$ and $x \in A$ such that $f(x) = 1$. Thus f maps A onto R and $A \cong \text{Ker } f \oplus R$. Since A is indecomposable, we have $A \cong R$, and rank $A = 1$. Thus we can assume that $I \subseteq M$.

We now have $F = M^{-1} \subset I^{-1}$. Let $v \in F$ and $x \in A$, and define $vx \in A''$ by $(vx)(f) = vf(x)$ for all $f \in A'$. Since $I^{-1}I \subset R$, vx is well-defined. This makes A'' into an F-module. Since $A \cong A''$, A is a torsion-free F-module of finite rank. Now F is a maximal valuation ring and hence it is a D-ring by Theorem 65. Therefore A is a direct sum of F-modules of rank 1. Thus, A is a direct sum of R-modules of rank 1, and hence rank $A = 1$.

This completes the proof that R has property D.

Examples: The following two examples show that there are Noetherian local rings and non-Noetherian quasi-local rings which have property D but are not maximal valuation rings, and thus are not integrally closed.

(1) Let k be a field and X an indeterminate over k. Let $F = k[[X]]$, the ring of formal power series in X with coefficients in k. Then F is a complete discrete valuation ring. Let R be the ring of those power series in F with no linear term. Then R is a complete, Noetherian, local domain with maximal ideal M generated by X^2 and X^3. F is the integral closure of R and is generated as an R-module by 1 and X. Furthermore $FM = M$. Thus R is a D-ring by Theorem 75. Since it is easily shown that every ideal of R can be generated by two elements we could also have applied Theorem 64 to show that R is a D-ring.

The next example is a generalization of the above construction to the non-Noetherian case.

(2) Let k be a field and X an indeterminate over k. Let G be the Abelian group $Z \times Z$, ordered lexicographically, where Z is the group of integers. Let F be the ring of all formal power series $\sum_{\alpha} b_{\alpha} X^{\alpha}$, where $b_{\alpha} \in k$, $\alpha \in G$, $\alpha \geq (0, 0)$, and the sets of exponents $\{\alpha\}$ are well-ordered. By [33, Ch. 2, Cor. Th. 8], F is a maximal valuation ring and the maximal ideal N of F is the set of power series with exponents $\alpha > (0, 0)$.

Let R be the subset of F consisting of those power series where the term $X^{(0, 1)}$ has zero coefficient. Then R is a quasi-local ring with maximal ideal M consisting of those power series with exponents $\alpha > (0, 1)$. We have $FM = M$, and F is generated as an R-module by the two elements $1 = X^{(0, 0)}$ and $X^{(0, 1)}$. Thus R is a D-ring by Theorem 75. R is not a maximal valuation ring since F is the integral closure of R, and $F \neq R$. F is a rank 2 valuation ring, and thus is not Noetherian. Hence R is not a Noetherian ring.

This chapter is devoted to proving Theorem 78 which is an important lemma for Theorem 83 of the next chapter. Theorem 78 states that an h-local domain with more than one maximal ideal is a D-ring if and only if it is a ring of type I. Thus, in particular, an h-local D-ring that is not a quasi-local ring is a Prüfer ring, and is necessarily integrally closed. Theorems 76 and 77 are lemmas for Theorem 78.

THEOREM 76. Let R be an h-local ring with more than two maximal ideals. Then R has an indecomposable, torsion-free R-module of rank 2.

Proof. Let M_1, M_2, M_3 be three distinct maximal ideals of R. Let A be a two-dimensional vector space over Q with basis u and v. Choose an element $b \in M_3$ such that $b \notin M_1$ and $b \notin M_2$, and let $z = (1/b)w + v \in A$. We define three R-submodules C_1, C_2, and C_3 of A as follows. Let $C_1 = Qu \oplus R_{M_1} v$; $C_2 = R_{M_2} u \oplus Qv$, and $C_3 = R_{M_3} u \oplus R_{M_3} z$. We let $C = C_1 \cap C_2 \cap C_3$ and we will show that C is an indecomposable, torsion-free R-module of rank 2.

We not that $z \in C$; for certainly $z \in C_1$ and C_3; and since $b \notin M_2$, we have $z \in C_2$ also. We will show that $C_i = C_{M_i}$ for $i = 1, 2, 3$. Since $C_{M_i} \subseteq (C_i)_{M_i} = C_i$, it will be sufficient to show that $C_i \subseteq C_{M_i}$ for all i.

115

(1) $\underline{C_1 = C_{M_1}}$.

Since $(R_{M_2} \cap R_{M_3})u \subset C$, we have $R_{M_1}(R_{M_2} \cap R_{M_3})u$
$\subset C_{M_1}$. But R is an h-local ring and hence $R_{M_1}(R_{M_2} \cap R_{M_3}) = Q$
by Theorem 22. Therefore, $Qu \subset C_{M_1}$. In particular, $(1/b)u \in C_{M_1}$,
and hence $v = z - (1/b)u \in C_{M_1}$. Thus $C_1 = Qu \oplus R_{M_1}v \subset C_{M_1}$.

(2) $\underline{C_2 = C_{M_2}}$.

Since $(R_{M_1} \cap R_{M_3})bv \subset C$, we have $R_{M_2}(R_{M_1} \cap R_{M_3})bv$
$\subset C_{M_2}$. But R is an h-local ring, and hence $R_{M_2}(R_{M_1} \cap R_{M_3})b =$
$Qb = Q$ by Theorem 22. Therefore $Qv \subset C_{M_2}$, and hence
$C_2 = R_{M_2}u \oplus Qv \subset C_{M_2}$.

(3) $\underline{C_3 = C_{M_3}}$.

Since u, z are in C, we have $C_3 = R_{M_3}u \oplus R_{M_3}z \subset C_{M_3}$.

Now C is a torsion-free R-module of rank 2, since it contains
u and z. We will assume that C is not indecomposable and arrive at
a contradiction. Suppose $C = D_1 \oplus D_2$, where D_1, D_2 are R-submodules of C of rank 1.

(4) $\underline{\text{We can assume that}}$ $(D_1)_{M_1} = Qu$ $\underline{\text{and}}$ $(D_2)_{M_2} = Qv$.

For we have $C_1 = C_{M_1} = (D_1)_{M_1} \oplus (D_2)_{M_1}$. Since C_1 is not
reduced, we can assume that $(D_1)_{M_1} \cong Q$. But Qu is the divisible
submodule of C_1, and thus $(D_1)_{M_1} = Qu$. A similar argument for C_2

shows that either $(D_1)_{M_1} = Qv$ or $(D_2)_{M_2} = Qv$. Since D_1 has rank 1, we must have $(D_2)_{M_2} = Qv$.

It follows from (4) that $D_1 \subset Qu$ and $D_2 \subset Qv$. Let $E_1 = (D_1)_{M_3}$ and $E_2 = (D_2)_{M_3}$. Thus $E_1 \subset Qu$, $E_2 \subset Qv$ and $C_3 = C_{M_3} = E_1 \oplus E_2$. Now C_3 is a free R_{M_3}-module of rank 2, and thus E_1 is a free R_{M_3}-module with generator x_1 and E_2 is a free R_{M_3}-module with generator x_2. We have $x_1 = q_1 u$ and $x_2 = q_2 v$, where $q_1, q_2 \in Q$.

Now $z \in C_3$, and hence there exists elements $d_1, d_2 \in R_{M_3}$ such that $(1/b)u + v = z = d_1 x_1 + d_2 x_2 = d_1 q_1 u + d_2 q_2 v$. Thus we have $1 = d_2 q_2$. Since $x_2 \in C_3$, there exist elements $a_1, a_2 \in R_{M_3}$ such that $q_2 v = x_2 = a_1 u + a_2 z = (a_1 + a_2/b)u + a_2 v$. Therefore $a_1 + a_2/b = 0$ and $a_2 = q_2$. Combining these results we have $0 = bd_2(a_1 + a_2/b) = bd_2 a_1 + d_2 q_2 = bd_2 a_1 + 1$. Hence $1/b = -d_2 a_1 \in R_{M_3}$. But $b \in M_3$, and so we have a contradiction. This contradiction proves that C is an indecomposable, torsion-free R-module of rank 2.

THEOREM 77. Let R be an h- local domain with more than one maximal ideal. If R is a D-ring, then inj. $\dim_R I = 1$ for every nonzero ideal I of R.

Proof. Assume that R is a D-ring and let I be a nonzero ideal of R. We will suppose that inj. $\dim_R I > 1$ and arrive at a contradiction. Since R is an h-local ring we have by Theorem 24 that inj. $\dim_R I = \sup_M$ inj. $\dim_{R_M} I_M$, where M ranges over all maximal ideals of R. Hence there exists a maximal ideal M such that inj. $\dim_{R_M} I_M > 1$. By Theorem A3 we have inj. dim$_R$ I$_M$ = inj. dim$_{R_M}$ I$_M$. Hence there exists a nonzero ideal J of R such that Ext$_R^2$(R/J, I$_M$) $\neq$ 0. Since Ext$_R^2$(R/J, I$_M$) $\cong$ Ext$_R^1$(J, I$_M$), there exists an exact sequence of R-modules

$$(1) \qquad\qquad 0 \to I_M \to A \xrightarrow{f} J \to 0$$

which is <u>not</u> a split exact sequence.

Now A is a torsion-free R-module of rank 2, and hence by property D, we have A = A$_1$ $\oplus$ A$_2$, where A$_i$ is a torsion-free R-module of rank 1. We can assume without loss of generality that the projection g of I$_M$ into A$_1$ is not zero. Now a homomorphism between two torsion-free modules of rank 1 is either zero or a mono-morphism. Thus g is a monomorphism and we have an exact sequence.

$$0 \to I_M \xrightarrow{g} A_1 \to B \to 0$$

where B is necessarily a torsion R-module. Applying the functor Hom$_R$($\cdot$, J) to this sequence we obtain the exact sequence:

$$0 \to \mathrm{Hom}_R(B, J) \to \mathrm{Hom}_R(A_1, J) \to \mathrm{Hom}_R(I_M, J).$$

Since B is a torsion R-module, we have Hom$_R$(B, J) = 0. To prove that Hom$_R$(A$_1$, J) = 0 it is sufficient to prove that Hom$_R$(I$_M$, J) = 0. Suppose that Hom$_R$(I$_M$, J) $\neq$ 0. Then we have a monomorphism of I$_M$ into J, and thus I$_M$ is isomorphic to an ideal of R. Now R$_M$ is a D-ring by Theorem 59, and thus R$_M$ is complete in the R$_M$-topology by Theorem 61. By Theorem 14, I$_M$ is a cotorsion R$_M$-module; and thus by Theorem 15, I$_M$ is a cotorsion R-module. But then by Theorem 14, R is complete in the R-topology. Since R is an h-local ring with more than one maximal ideal, its completion is a direct product of rings and can not be an integral domain by Theorem 22(5). This contradiction shows that $\text{Hom}_R(I_M, J) = 0$, and hence $\text{Hom}_R(A_1, J) = 0$ also.

If we now return to exact sequence (1) we see that $\text{Hom}_R(A_1, J) = 0$ implies that $f(A_1) = 0$. Thus $A_1 \subseteq \text{Ker } f = I_M$. It follows that $f(A_2) = J$, and hence $I_M + A_2 = A$. Since A_2 and J are both torsion-free modules of rank 1, f is a monomorphism on A_2. Therefore, $A_2 \cap \text{Ker } f = 0$; that is, $A_2 \cap I_M = 0$. Thus $A = A_2 \oplus I_M$, which shows that (1) is a split exact sequence. This contradiction proves that inj. $\dim_R I = 1$.

DEFINITION. R is said to be a <u>ring of type I</u> if R is an integral domain satisfying the following conditions:

(1) R has exactly two maximal ideals M_1 and M_2.

(2) $M_1 \cap M_2$ does <u>not</u> contain a nonzero prime ideal of R.

(3) R_{M_1} and R_{M_2} are maximal valuation rings.

Of course a ring of type I is an h-local ring and a Prüfer ring. By Theorem 52 an equivalent defintion is that $R = V_1 \cap V_2$, where V_1 and V_2 are maximal valuation rings (not fields) with the same quotient field Q; and if W is a valuation ring contained in but not equal to Q, then W contains at most one of the rings V_1 and V_2.

It is a consequence of Theorems 64 and 78 that there are no Noetherian rings of type I. That rings of type I do exist, however, can be seen from the following example due to Barbara Osofsky.

Example. Let V be the ring of formal power series in an indeterminate X with coefficients in the complex numbers $\mathbb{C}$, and exponents well-ordered subsets of non-negative rational numbers; and let Q be the quotient field of V. There exists an automorphism σ of Q that keeps the elements of $\mathbb{C}$ fixed and sends X into $1 - X$. Let $W = \sigma(V)$, and let $R = V \cap W$. Then R is a ring of type I.

The following theorem shows the key role that rings of type I play in the theory of D-rings. We will sharpen this result further in later theorems.

THEOREM 78. Let R be an h-local domain with more than one maximal ideal. Then R is a D-ring if and only if R is a ring of type I. In this case, R is a Prüfer ring.

Proof. Suppose that R is a D-ring. By Theorem 76, R has exactly two distinct maximal ideals M_1 and M_2. By Theorem 59, R_{M_1} is a D-ring; and hence by Theorem 61, R_{M_1} is complete in the R_{M_1}-topology. Let J be a nonzero ideal of R_{M_1} and set $I = J \cap R$. By Theorem 77, $\text{inj.dim}_R I = 1$; and by Theorem 24,

$\text{inj.dim}_R I \geq \text{inj.dim}_{R_{M_1}} I_{M_1}$. Since $I_{M_1} = J$, we see that

$\text{inj.dim}_{R_{M_1}} J = 1$. Thus R_{M_1} is a complete domain such that every nonzero ideal has injective dimension 1 and hence R_{M_1} is a maximal valuation ring by Theorem 51(3). In similar fashion we see that R_{M_2} is a maximal valuation ring. Thus we have proved that R is a ring of type I.

Conversely, assume that R is a ring of type I. Before proving that R is a D-ring we will establish a number of properties of torsion-free modules over a ring of type I that we will need in the proof.

(1) If B is a torsion-free R-module of finite rank, then $\text{inj.dim}_R B \leq 1$.

By Theorem 24 we have $\text{inj.dim}_R B \leq \sup_i \text{inj.dim}_{R_{M_i}} B_{M_i}$;

and by Theorem 51 we have $\text{inj.dim}_{R_{M_i}} B_{M_i} \leq 1$ for $i = 1, 2$.

(2) If B is a reduced, torsion-free R-module of rank 1, then B is isomorphic to an ideal of R_{M_1} or R_{M_2} if and only if $B = B_{M_1}$ or $B = B_{M_2}$. These cases are mutually exclusive. B is a cotorsion R-module if and only if one of these two cases occurs. Otherwise, B is isomorphic to an ideal of R and is not a cotorsion R-module.

We will assume that B is a proper R-submodule of Q and that B is not isomorphic to an ideal of R. If $B_{M_1} \neq Q$ and $B_{M_2} \neq Q$, then, since R_{M_1} and R_{M_2} are valuation rings, there exist elements r and t in R such that $rB_{M_1} \subseteq R_{M_1}$ and $tB_{M_2} \subseteq R_{M_2}$. But then $rtB \subseteq R_{M_1} \cap R_{M_2} = R$, and B is isomorphic to an ideal of R. This contradiction shows that we can assume that $B_{M_1} = Q$. As a result we have $B = B_{M_1} \cap B_{M_2} = Q \cap B_{M_2} = B_{M_2}$. Since R_{M_2} is a valuation ring, B is thus isomorphic to an ideal of R_{M_2}. In this case by Theorems 14 and 15, B is a cotorsion R-module. On the other hand if B is isomorphic to an ideal of R, and B is a cotorsion R-module, then R is a cotorsion R-module. But then R is complete by Theorem 15, and since R is an h-local ring with more than one maximal ideal, this contradicts Theorem 22(5). Hence B is not a cotorsion R-module in this case.

(3) If C is a torsion-free, cotorsion R-module of finite rank, and B is any torsion-free R-module, then $\text{Ext}_R^1(B, C) = 0$.

From exact sequence (II) for B we derive the exact sequence:

$$\text{Ext}_R^1(Q \otimes_R B, C) \to \text{Ext}_R^1(B, C) \to \text{Ext}_R^2(K \otimes_R B, C) \ .$$

The first term of this sequence is zero since C is a cotorsion R-module, and the last term is zero since $\text{inj.dim}_R C = 1$ by (1). Thus $\text{Ext}_R^1(B, C) = 0$.

(4) *If* A *is a torsion-free* R-*module, and* C *is a pure, cotorsion submodule of* A *of finite rank, then* C *is a direct summand of* A.

If we let $B = A/C$, then B is torsion-free, since C is pure in A. But $\text{Ext}_R^1(B, C) = 0$ by (3), and since $\text{Ext}_R^1(B, C)$ is the set of equivalence classes of extensions of C by B, it follows that C is a direct summand of A.

(5) *If* A *is a torsion-free* R-*module of finite* rank > 1, *and if* f *is a nonzero element of* $\text{Hom}_R(A, R)$, *then* $\text{Ker } f$ *is a proper, nonzero direct summand of* A.

Let $B = \text{Ker } f$ and $I = \text{Im } f$; then I is an ideal of R and B is a proper, nonzero submodule of A. Since $\text{inj.dim}_R B \leq 1$ by (1), we have $\text{Ext}_R^2(R/I, B) = 0$. But $\text{Ext}_R^2(R/I, B)$ is isomorphic to $\text{Ext}_R^1(I, B)$, and thus $\text{Ext}_R^1(I, B) = 0$. Since A is an extension of B by I, it follows that B is a direct summand of A.

(6) *If* A *is a torsion-free* R-*module of finite rank* > 1, *and if there is a homomorphism of* A *onto* Q, *then* A *has a proper, nonzero direct summand*.

We can assume that A is reduced and we let g be a homomorphism of A onto Q with $B = \text{Ker } g$. We will not prove that B is a direct summand of A, but we will find another submodule of A that is. Since R_{M_1} is a maximal valuation ring, $\text{Ext}_{R_{M_1}}^1(Q, B_{M_1}) = 0$ by

Theorem 51. Thus we see that $A_{M_1} = B_{M_1} \oplus Q_1$, where Q_1 is an R-submodule of A_{M_1} and $Q_1 \cong Q$. We can assume that $A \subset A_{M_1}$, and we set $C = Q_1 \cap A$. Since A_{M_1} is an essential extension of A, C is a submodule of A of rank 1. C is pure in A because Q_1 is pure in A_{M_1}. Now $C_{M_1} = Q_1 \cap A_{M_1} = Q_1$ and $C_{M_2} \subset Q_1$. Therefore, $C = C_{M_1} \cap C_{M_2} = Q_1 \cap C_{M_2} = C_{M_2}$; and since C is reduced, it follows from (2) that C is a cotorsion R-module. By (4), C is a direct summand of A.

(7) If C is a torsion-free, cotorsion R-module of rank 1, then $\mathrm{Ext}_R^1(C, R)$ is a torsion R-module.

By (2) we can assume that C is an ideal of R_{M_1}, and thus we have an exact sequence:

$$\mathrm{Ext}_R^1(R_{M_1}, R) \to \mathrm{Ext}_R^1(C, R) \to \mathrm{Ext}_R^2(R_{M_1}/C, R) .$$

Since $\mathrm{inj.dim}_R R = 1$ by (1), the last term of this sequence is zero. Hence $\mathrm{Ext}_R^1(C, R)$ is a homomorphic image of $\mathrm{Ext}_R^1(R_{M_1}, R)$, and to prove (7) it will be sufficient to prove that $\mathrm{Ext}_R^1(R_{M_1}, R)$ is a torsion R-module.

Because R is an h-local ring, we have $R_{M_1} \oplus R_{M_2} = Q$ by Theorem 22(2). Since $R_{M_1} \cap R_{M_2} = R$, we have an exact sequence:

$$0 \to R \to R_{M_1} \oplus R_{M_2} \to Q \to 0 .$$

Now R_{M_1} and R_{M_2} are maximal valuation rings and hence are complete rings. Thus by Theorem 22(5), $R_{M_1} \oplus R_{M_2}$ is isomorphic to H, the completion of R. The preceding exact sequence then shows that $H/R \cong Q$. Since $H/R \cong \mathrm{Ext}_R^1(Q, R)$ by Theorem 10, we have $\mathrm{Ext}_R^1(Q, R) \cong Q$. By (2), R has no nonzero cotorsion ideals and thus

$\mathrm{Hom}_R(R_{M_1} \oplus R_{M_2}, R) = 0$. Thus if we apply the functor $\mathrm{Hom}_R(\cdot, R)$ to the preceding exact sequence, we obtain the exact sequence:

$$0 \to R \to Q \to \mathrm{Ext}^1_R(R_{M_1} \oplus R_{M_2}, R) \to 0 .$$

Therefore, the last term of this sequence is isomorphic to K and hence is a torsion R-module. Since $\mathrm{Ext}^1_R(R_{M_1}, R)$ is a direct summand of this last term, it is also a torsion R-module.

We are now ready to prove that R is a D-ring. Suppose that R is not a D-ring. Then there exists an indecomposable, torsion-free R-module A of finite rank $n > 1$, and we can assume that every non-zero, torsion-free R-module of rank $< n$ is a direct sum of R-modules of rank 1. Clearly A is a reduced R-module. Let B be a pure submodule of A of rank $n-1$, and let $C = A/B$. Then C is a torsion-free R-module of rank 1, and $B = B_1 \oplus \ldots \oplus B_{n-1}$, where B_i is a submodule of B of rank 1 for $i = 1, 2, \ldots, n-1$.

Since B_1 is pure in B, and B is pure in A, B_1 is a pure submodule of A. By (4), B_1 is not a cotorsion module, and hence by (2), B_1 is isomorphic to an ideal of R. Thus $\mathrm{Hom}_R(B_1, R) \neq 0$, and hence $\mathrm{Hom}_R(B, R) \neq 0$. We have an exact sequence:

$$\mathrm{Hom}_R(A, R) \to \mathrm{Hom}_R(B, R) \to \mathrm{Ext}^1_R(C, R) .$$

Since $\mathrm{Hom}_R(A, R) = 0$ by (5), we conclude that $\mathrm{Hom}_R(B, R)$ is isomorphic to a nonzero submodule of $\mathrm{Ext}^1_R(C, R)$, and thus $\mathrm{Ext}^1_R(C, R)$ is not a torsion R-module. But by (5) and (6), C is not isomorphic to either Q or an ideal of R, and hence C is a cotorsion module by (2). Thus by (7), $\mathrm{Ext}^1_R(C, R)$ is a torsion module, and this contradiction proves that R is a D-ring.

13

RINGS OF TYPE I

The main result of this chapter is Theorem 83, where we prove that a domain is a ring of type I if and only if it is a D-ring with a remote quotient field. Theorem 79 clarifies the meaning of a remote quotient field. Theorem 80 describes a property of a direct sum decomposition; and its corollary, Theorem 81, is a lemma for Theorem 83. However, the chief lemma for Theorem 83 is Theorem 82 in which some ideal theoretic properties of a D-ring are examined. Theorem 84 is again a theorem about direct sum decompositions, and is a lemma for Theorem 85 which gives an alternative characterization of a ring of type I as a noncomplete D-ring.

DEFINITION. The domain R is said to have a <u>remote quotient field</u>, if there exists an R-submodule S of Q, $S \neq Q$, with $S^{-1} = 0$; that is if there exists a reduced torsion-free R-module of rank 1 that is not torsionless.

Clearly a torsionless ring does <u>not</u> have a remote quotient field. On the other hand if a domain does not have a remote quotient field, then, by Theorem 42, it is a torsionless ring if and only if it is complete. Thus we see that having a remote quotient field is not a great restriction on a domain. In fact a Noetherian integral domain has a remote quotient field if and only if it is <u>not</u> a local ring of Krull dimension 1 whose integral closure is a valuation ring that is a finitely generated module. It is true in general that if a domain has a remote quotient field, then it is quite far from being a valuation ring, as we see in the following theorem.

THEOREM 79. The following statements are equivalent:

(1) R has a remote quotient field.

(2) There exists a valuation ring V, $R \subset V \subsetneq Q$, such that $V^{-1} = 0$.

(3) If V is a valuation ring, $R \subset V \subset Q$, then $V^{-1} = 0$.

Proof. Clearly we have $(3) \Rightarrow (2) \Rightarrow (1)$. We will assume (1) and prove (3). Suppose that there exists a valuation ring V, $R \subset V \subset Q$, such that $V^{-1} \neq 0$. Then, of course, we have $V \neq Q$. By (1) there exists an R-module S, $S \subsetneq Q$, such that $S^{-1} = 0$. Choose $r \in R$, $r \neq 0$, such that $rV \subset R$. If $VS = Q$, then $Q = rQ = rVS \subset S$; and this contradiction shows that $VS \neq Q$. Since the V-submodules of Q are linearly ordered, there exists $t \in R$, $t \neq 0$, such that $t(VS) \subset V$. Then $rtS \subset rtVS \subset rV \subset R$, and we have $S^{-1} \neq 0$. This contradiction shows that $V^{-1} = 0$ for every valuation ring V such that $R \subset V \subset Q$.

THEOREM 80. Let S be an R-submodule of Q such that $S^{-1} = 0$. Let I be a nonzero ideal of R and let A be an extension of S by I:

$$0 \to S \to A \to I \to 0 .$$

Then A is a decomposable R-module if and only if S is one of the component direct summands of A.

Proof. Suppose that A has a nontrivial decomposition:
$A = A_1 \oplus A_2$. Let $\pi_i : S \to A_i$ be the canonical projections of S into
A_i for $i = 1, 2$. Since A_i is torsion-free and rank $S = 1 = \text{rank } A_i$,
we see that either $\pi_i = 0$ or π_i is a monomorphism for $i = 1, 2$.

Let $f : A \to I$ be the canonical map coming from the fact that A
is an extension of S by I. Suppose that both $\pi_i \neq 0$ and $\pi_2 \neq 0$. Then
there is a monomorphism $A_i^{-1} \to S^{-1}$ for $i = 1, 2$. Hence $A_1^{-1} = 0 = A_2^{-1}$ and $f | A_i = 0$ for $i = 1, 2$. But then $f = 0$. This contradiction
shows that without loss of generality we can assume that $\pi_2 = 0$.

We thus have $S \subset A_1$. Since $S = \text{Ker } f$, we have an induced
monomorphism $A_1/S \to I$. But A_1/S is a torsion R-module and I is
torsion-free. Therefore $S = A_1$.

Compare the following theorem with Theorem 77 and statement
(1) of Theorem 78!

THEOREM 81. Let R be a D-ring and let S be an
R-submodule of Q such that $S^{-1} = 0$. Then inj. $\dim_R S \leq 1$.

Proof. Let I be a nonzero ideal of R and consider an exact
sequence of the form
$$0 \to S \to A \to I \to 0.$$

Then A is torsion-free of rank 2, and so by property D, A is decomposable. Hence by Theorem 80, the above sequence splits. Therefore,
$\text{Ext}_R^1(I, S) = 0$. Since $\text{Ext}_R^2(R/I, S) \cong \text{Ext}_R^1(I, S) = 0$, it follows that
inj. $\dim_R S \leq 1$.

THEOREM 82. Let R be a D-ring and let S be an R-module,
$R \subset S \subset Q$ such that $S^{-1} = 0$. Let I be an ideal of R and b an ele-

ment of R such that $(I: b)$ is contained in the Jacobson radical of R. Then

$$I = (Sa \cap I) + (I: b)b$$

for every nonzero element $a \in R$.

Proof. Let X be a two-dimensional vector space over Q with basis x and y. Choose $a \in R$, $a \neq 0$, and define $A = Sx + Iy + R(\frac{1}{a}x + by)$; then A is an R-submodule of X. Clearly A is a torsion-free R-module of rank 2, and hence there exists a nontrivial decomposition $A = A_1 \oplus A_2$. Let $\theta: A \to A_1$ be the canonical projection of A onto A_1 with kernel A_2.

Let $w \in A$; then $w = q_1 x + q_2 y$, where $q_1, q_2 \in Q$. But $w = sx + cy + r(\frac{1}{a}x + by)$, where $s \in S$, $c \in I$, and $r \in R$. Hence we have

$$(1) \qquad w = (s + \frac{r}{a})x + (c + rb)y .$$

Thus $q_1 = s + \frac{r}{a}$ and $q_2 = c + rb$. Hence $q_2 \in I + Rb$. Now $\theta(x) = z \in A_1$, and $z = q_3 x + q_4 y$, where $q_3, q_4 \in Q$. Suppose that $q \in A$, where $q \in Q$, $q \neq 0$. Then $q = \frac{m}{n}$, where $m, n \in R$ and $n \neq 0$. We have $n\theta(qx) = \theta(nqx) = \theta(mx) = m\theta(x)$. Hence in X we have $\theta(qx) = q\theta(x)$.

Let $t \in S$; then $\theta(tx) = t\theta(x) = tz = tq_3 x + tq_4 y \in A$. Then $tq_4 \in I + Rb \subset R$, and so $q_4 S \subset R$. Since $S^{-1} = 0$, we have $q_4 = 0$. Therefore $\theta(x) = z = q_3 x$. Now $x = z + w$, where $z \in A_1$ and $w \in A_2$. Hence $w = x - z = x - q_3 x = (1 - q_3)x$.

Suppose that $q_3 \neq 0$ and $1 - q_3 \neq 0$. Then there exist elements $r_1, r_2 \in R$, $r_1 \neq 0$ and $r_2 \neq 0$, such that $r_1 q_3 = r_2(1 - q_3)$. We would then have $r_1 z = r_1 q_3 x = r_2(1 - q_3)x = r_2 w$. Thus $r_1 z$ would be a nonzero element of $A_1 \cap A_2 = 0$. This contradiction shows that either

$q_3 = 0$ or $1 - q_3 = 0$. Without loss of generality we can assume that $1 - q_3 = 0$. Hence $q_3 = 1$ and so $\theta(x) = x$.

We have the following equation

$$(2) \qquad A \cap Qx = (S + \frac{I:b}{a})x \ .$$

For let $w \in A \cap Qx$. Then by (1), w is of the form $w = (s + \frac{r}{a})x$; and $c + rb = 0$. Hence $r \in (I:b)$ and so $w \in (S + \frac{(I:b)}{a})x$. Conversely, by the definition of A we have $Sx \subset A \cap Qx$. Let $r \in (I:b)$; then $\frac{r}{a} x = r(\frac{1}{a} x + by) - rby \in A$. Hence $(\frac{I:b}{a})x \subset A \cap Qx$. Hence we have proved that equation (2) is valid.

We will next prove the following equation:

$$(3) \qquad A_1 = (S + \frac{(I:b)}{a})x \ .$$

For if $qx \in A$, we have seen that $\theta(qx) = q\theta(x) = qx$. Hence $A \cap Qx \subset A_1$. Conversely, let $w \in A_1$. Since rank $A_1 = 1$ and $Rx \subset A_1$, there exists $r \in R$, $r \neq 0$, such that $rw \in Rx$. Therefore, the coefficient of y in w is 0, and $w \in A \cap Qx$. Hence $A_1 = A \cap Qx = (S + \frac{(I:b)}{a})x$. (Parenthetically, we observe that we have shown that $(S + \frac{(I:b)}{a})x$ is one of the components in every direct sum decomposition of A.)

Now $x + aby = a(\frac{1}{a}x + by) \in A$. Since $x \in A$, it follows that $aby \in A$. Thus $\theta(aby) = ux$, where $u \in (S + \frac{(I:b)}{a})$. Hence we have $\theta(x + aby) = x + ux = (1 + u)x$. Therefore, $a\theta(\frac{1}{a}x + by) = \theta(x + aby) = (1 + u)x$, and so $\theta(\frac{1}{a}x + by) = \frac{1 + u}{a} x$.

Let $c \in I$; then $ab\theta(cy) = \theta(abcy) = c\theta(aby) = cux$. Hence $\theta(cy) = \frac{cu}{ab} x$.

We next prove the following inequality:

(4) $(1 + u) + I \frac{u}{b} \subseteq [Sa + (I : b)].$

For $\frac{1+u}{a} x = \theta(\frac{1}{a} x + by) \in A_1 = (S + \frac{I : b}{a})x.$ Hence $1 + u \in [Sa + (I : b)].$
We also have $\frac{Iu}{ab} x = \theta(Iy) \subseteq A_1 = (S + \frac{(I : b)}{a})x.$ Therefore,
$\frac{Iu}{b} \subseteq [Sa + (I : b)]$ also. This establishes (4).

Since $1 + u \in [Sa + (I : b)]$, we have $1 + u = sa + d$, where $s \in S$
and $d \in (I : b)$. Thus $u = sa + (d - 1)$. Let $c \in I$. Then since
$\frac{Iu}{b} \subseteq [Sa + (I : b)]$, we have $c\frac{u}{b} = ta + e$, where $t \in S$ and $e \in (I : b)$.
Therefore, $ta + e = \frac{csa + c(d - 1)}{b}$, and so $tab + eb = csa + c(d - 1)$.
Hence $(d - 1)c = tab + eb - csa = (bt - cs)a + eb.$

Since $d \in (I : b)$ which is contained in the Jacobson radical of R,
$(d - 1)$ is a unit in R. Thus $c = [(d - 1)^{-1}(bt - cs)]a + (d - 1)^{-1}eb$. Let
$e' = (d - 1)^{-1}e$ and $v = (d - 1)^{-1}(bt - cs)$. Then

$$c = va + e'b$$

where $v \in S$ and $e' \in (I : b)$. Since $va = c - e'b \in I + (I : b)b = I$, it follows
that $c \in (Sa \cap I) + (I : b)b$. Since c was an arbitrary element of I, we
have $I \subseteq (Sa \cap I) + (I : b)b$ which establishes the theorem.

THEOREM 83. R is a ring of type I if and only if R is a
D-ring with a remote quotient field.

Proof. Suppose that R is a ring of type I. Then by Theorem
78, R is a D-ring. We must show that it has a remote quotient field Q.
Let M_1 and M_2 be the two distinct maximal ideals of R. Since R is
h-local, it follows from Theorem 22(2) that $R_{M_1} + R_{M_2} = Q$. Suppose
that $R_{M_1}^{-1} \neq 0$ and $R_{M_2}^{-1} \neq 0$. Let $I = R_{M_1}^{-1} \cap R_{M_2}^{-1}$. Then $I \neq 0$, and

so $Q = IQ = I(R_{M_1} + R_{M_2}) = IR_{M_1} + IR_{M_2} \subseteq R$. This contradiction shows that either $R_{M_1}^{-1} = 0$ or $R_{M_2}^{-1} = 0$. Thus R has a remote quotient field.

We will assume from now on that R is a D-ring with a remote quotient field. We will prove a sequence of statements about R that will eventually show that R is a ring of type I.

(1) <u>Let</u> F <u>be the integral closure of</u> R. <u>Then</u> F <u>is either a maximal valuation ring or a ring of type</u> I.

We may assume that F is not a maximal valuation ring. We must then prove that F is a ring of type I. If F is a valuation ring, $R \subseteq F \subsetneq Q$, then F is a D-ring by Theorem 59, and so F is a maximal valuation ring by Theorem 65. Thus F is not a valuation ring.

Let ζ be the collection of valuation rings V such that $R \subseteq V \subsetneq Q$ and such that V is minimal with respect to this property. Since F is the intersection of the valuation rings containing R, we can apply Zorn's Lemma and obtain $F = \bigcap_{V \in \zeta} V$.

Since F is not a valuation ring, ζ has at least two distinct elements V and W. By minimality, $V \not\subseteq W$ and $W \not\subseteq V$. Let $m(V)$ and $m(W)$ be the maximal ideals of V and W, respectively. Let $S = V \cap W$; $M = m(V) \cap S$; and $N = m(W) \cap S$. Then, using Theorem 52 and the preceding remarks, we see that S is a Prüfer ring such that $S_M = V$ and $S_N = W$ are maximal valuation rings. We will prove that S is an h-local ring. According to Theorem 54 to do this it will be necessary and sufficient to prove that $\operatorname{inj.dim}_S S = 1$.

Suppose that S is not an h-local ring. Then there exists a nonzero prime ideal P of S such that $P \subset M \cap N$. Since $V = S_M \subset S_P$, we see that S_P is a valuation ring with maximal ideal $m(S_P) = PS_P$ and $PS_P \subset V$. Similarly $PS_P \subset W$. Hence $PS_P \subset V \cap W = S$, and thus $P = PS_P \cap S = PS_P$. Therefore, there exists a nonzero element $r \in R$ such that $rS_P \subset S$. Since R has a remote quotient field, we have by Theorem 79 that $S_P^{-1} = 0$. Since $S^{-1} \subset \frac{1}{r} S_P^{-1} = 0$, it follows from Theorem 81 that $\mathrm{inj.\,dim}_R S = 1$. We still have to show that S has injective dimension 1 over S.

Now there exists an exact sequence of R-modules:

$$0 \to S \to V \oplus W \to V + W \to 0.$$

Since $V + W \subset S_P$, $V + W$ is reduced. Also V and W being maximal valuation rings are cotorsion modules over themselves, and hence cotorsion modules over R as well by Theorem 15. Therefore, S is a cotorsion R-module by Theorem 1 and $\mathrm{Ext}_R^1(Q, S) = 0$. Let I be a nonzero ideal of S. Since $\mathrm{inj.\,dim}_R S = 1$, we see that $\mathrm{Ext}_R^2(Q/I, S) = 0$. It follows from the exact sequence

$$\mathrm{Ext}_R^1(Q, S) \to \mathrm{Ext}_R^1(I, S) \to \mathrm{Ext}_R^2(Q/I, S) \ ,$$

that $\mathrm{Ext}_R^1(I, S) = 0$. Consider an exact sequence of S-modules and S-homomorphisms of the form:

$$0 \to S \to A \to I \to 0.$$

Since $\mathrm{Ext}_R^1(I, S) = 0$, this splits over R, and hence by Theorem 58 it splits over S as well.

Hence $\mathrm{Ext}_S^1(I,S) = 0$. Since $\mathrm{Ext}_S^2(S/I,S) \cong \mathrm{Ext}_S^1(I,S) = 0$, we have inj. dim$_S$ S = 1. We have thus shown that S is an h-local ring. Thus S is a ring of type I.

We will assume that $S \neq F$, and arrive at a contradiction which will prove our assertion that F is a ring of type I. Since $S \neq F$, $\mathcal{C}$ has a distinct third element U such that if $T = U \cap V \cap W$, then $T \neq S$. By the independence of valuations of Theorem 52, T has exactly three maximal ideals: $m(U) \cap T$, $m(V) \cap T$, and $m(W) \cap T$, and the localizations of T with respect to these ideals are U, V, W, respectively.

Since T is a D-ring by Theorem 59, we have by Theorem 76 that T is not an h-local ring. Hence there exists a nonzero prime ideal P^* of T which (without loss of generality) we can assume is contained in $(m(V) \cap T) \cap (m(W) \cap T)$. Then $V = T_{m(V) \cap T} \subset T_{P^*}$, and hence $P^*T_{P^*} \subset m(V)$. Similarly we have $W = T_{m(W) \cap T} \subset T_{P^*}$, and hence $P^*T_{P^*} \subset m(W)$. Therefore, $P^*T_{P^*}$ is a nonzero prime ideal of $S = V \cap W$ that is contained in $m(V) \cap m(W) = M \cap N$. This is a contradiction, since S is an h-local ring. Thus $F = S$ and F is a ring of type I.

(2) **Either** R **is a quasi-local ring, or** R **is a ring of type I.**

Suppose that R is not a quasi-local ring. Then the integral closure F of R can not be a valuation ring. Hence by (1), F is a ring of type I. Let N_1 and N_2 be the two distinct maximal ideals of F, and let $M_1 = N_1 \cap R$ and $M_2 = N_2 \cap R$. Then M_1 and M_2 are the only maximal ideals of R, and $M_1 \neq M_2$.

If R is an h-local ring, then by Theorem 78, R is a ring of type I. Hence we will assume that R is not an h-local ring and arrive at a contradiction. Thus we can assume that R has a nonzero prime ideal P such that $P \subset M_1 \cap M_2$. Since F is the integral closure of R, there exists a nonzero prime ideal P^* of F such that $P^* \cap R = P$. Since F is an h-local ring, we can assume that $P^* \subset N_1$ and $P^* \not\subset N_2$.

Let I be an ideal of F containing P^*. Then $I \subset N_1$ and $I \not\subset N_2$. F has only the two maximal ideals N_1 and N_2, and so by Theorem 25 we have $F_{N_1} I \cap F = I$. Since F_{N_1} is a valuation ring it follows that the ideals of F that contains P^* are linearly ordered. Therefore, F/P^* is a valuation ring.

Since F is integral over R and $P^* \cap R = P$, we see that F/P^* is integral over R/P. Since F/P^* is a valuation ring it follows that R/P is a quasi-local ring. But R/P has two distinct maximal ideals M_1/P and M_2/P. This contradiction shows that R is an h-local ring, and hence is a ring of type I. This proves (2).

According to (2) we can assume that R is a quasi-local D-ring with a remote quotient field. We will call such a ring a θ-ring. To complete the proof of Theorem 83 it will be necessary and sufficient to prove that θ-rings do not exist.

(3)   <u>Let R be a θ-ring, with maximal ideal M. Then the integral closure F of R is a maximal valuation ring with maximal ideal N and $R/M \cong F/\mathbf{N}$, and $F^{-1} = 0$.</u>

Suppose that F is not a maximal valuation ring. Then by (1) F is a ring of type I. Let N_1 and N_2 be the two distinct maximal ideals of F. We have $F_{N_1} \cap F_{N_2} = F$; and by Theorem 23(2) we have $F_{N_1} + F_{N_2} = Q$, the quotient field of R. Hence whe have an exact sequence:

$$0 \to F \to F_{N_1} \oplus F_{N_2} \to Q \to 0 .$$

From this we derive the exact sequence:

$$\operatorname{Hom}_R(F_{N_1}, R) \oplus \operatorname{Hom}_R(F_{N_2}, R) \to \operatorname{Hom}_R(F, R) \to \operatorname{Ext}_R^1(Q, R).$$

Since R is a quasi-local D-ring we have by Theorem 61 that R is complete in the R-topology. Thus R is a cotorsion R-module by Theorem 9 and so $\operatorname{Ext}_R^1(Q, R) = 0$. Since R has a remote quotient field, and F_{N_i} is a valuation ring, we have by Theorem 79 that $\operatorname{Hom}_R(F_{N_i}, R) = 0$ for $i = 1, 2$. Hence from the preceding exact sequence we see that $\operatorname{Hom}_R(F, R) = 0$; that is $F^{-1} = 0$.

Let $x \in N_1$, $x \notin N_2$, and let $R_1 = R[x]$. Since x is integral over R, R_1 is a finitely generated R-module. Hence $R_1^{-1} \neq 0$; and since $F^{-1} = 0$, we have $R_1 \neq F$. However, F is the integral closure of R_1, and thus R_1 is not integrally closed. Now R_1 has a remote quotient field. For if there exists an element $q \in Q$, $q \neq 0$, such that $qF \subset R_1$, then $(R_1^{-1} q)F \subset R$, and this contradicts $F^{-1} = 0$. By Theorem 59, R_1 is a D-ring. Hence by (2), R_1 is either a quasi-local ring or a ring of type I. Since rings of type I are integrally closed, it follows that R_1 is a quasi-local ring with maximal ideal P.

Since x is not invertible in F, we have $x \in P$. But F is the integral closure of R_1, and thus $N_1 \cap R_1 = P = N_2 \cap R_1$. Therefore,

$x \in N_2$. This contradiction shows that F is a maximal valuation ring with maximal ideal N.

Since R has a remote quotient field and F is a valuation ring, we have $F^{-1} = 0$ by Theorem 79. Now $MF \subset N$, and thus we must have $N^{-1} = 0$ also. Thus $\text{inj. dim}_R N = 1$ by Theorem 81. It follows that Q/N is an injective R-module.

Now Q/N is an indecomposable R-module. For N is a cotorsion F-module, and hence by Theorem 15, N is a cotorsion R-module. Therefore, Q/N is an indecomposable injective R-module by Theorem 16.

Now F/N is contained in the socle of Q/N considered as an R-module. However, since Q/N is an indecomposable injective R-module, the socle of Q/N is a simple R-module by Theorem 38. Hence we have $F/N \cong R/M$.

(4) <u>Let R be a θ-ring with maximal ideal</u> M. <u>Then</u> $M^{-1} \neq R$.

By (3) the integral closure F of R is a maximal valuation ring with maximal ideal N and $F^{-1} = 0$. Let $c \neq 0 \in M$ and define

$$L_c = \{ \bigcap Fb \mid b \in R, \, b \notin Rc \}.$$

Clearly, L_c is an ideal of F, and thus the set $\{ L_c \mid c \neq 0 \in M \}$ is a linearly ordered set. Since R is not a valuation ring, the principal ideals of R are not linearly ordered. Hence there exists an element $c \in M$, $c \neq 0$, such that $Rc \neq L_c \cap R$.

Now $Rc \subsetneq L_c \cap R$. For if $b \in R$ and $b \notin Rc$, then we can apply Theorem 82 to obtain

$$Rc = (Fb \cap Rc) + (Rc : b)b.$$

Thus $Rc \subset Fb$, and so $Rc \subsetneq L_c \cap R$.

Choose any element $b \in L_c \cap R$ such that $b \notin Rc$. Then Fb is one of the terms of the intersection L_c, and hence $L_c \subseteq Fb$. But $b \in L_c$, which is an ideal of F, and thus $Fb \subseteq L_c$. Therefore, $L_c = Fb$.

Choose $d \in M$, $d \neq 0$; then $db \in L_c \cap R$. If $db \notin Rc$, then as before we have $L_c = Fbd$. Hence $Fbd = Fb$, and thus $Fd = F$. But $d \in M \subseteq N$ and this is a contradiction. Therefore, $db \in Rc$; and this proves that $Mb \subseteq Rc$. Let $x = b/c \in Q$. Then $x \notin R$, but $Mx \subseteq R$. Hence $x \in M^{-1}$, $x \notin R$, proving that $M^{-1} \neq R$.

(5) <u>Let R be a θ-ring with maximal ideal M. Then M is a principal ideal of R.</u>

Suppose that M is not a principal ideal of R. Then M^{-1} is a ring, and by (4), M^{-1} properly contains R. Let S be the integral closure of M^{-1}. Then the integral closure F of R is contained in S. Since F is a maximal valuation ring, S is also a valuation ring. Thus M^{-1} is a quasi-local ring with maximal ideal P. Let B be the maximal ideal of S; then $P = B \cap M^{-1}$.

Since $F \subseteq S$, we have $B \subseteq N$, where N is the maximal ideal of F. Since M is not a principal ideal of R, we have $M^{-1}M = M$, and thus M is an ideal of M^{-1}. Therefore, $M \subseteq P$, and hence $P \cap R = M$. Thus $B \cap R = B \cap M^{-1} \cap R = P \cap R = M$. Hence the integral domain F/B is integral over the field R/M. But then F/B is a field, and thus $B = N$. This proves that $F = S$, the integral closure of M^{-1}, and hence $M^{-1} \subseteq F$.

By (3), we have $R/M \cong F/N$, and hence $F = R + N$. This shows that $M^{-1} = R + P$. For if $x \in M^{-1}$, then $x \in F$, and thus $x = r + y$, where $r \in R$ and $y \in N$. Hence $y = x - r \in M^{-1} \cap N = P$.

Thus $x \in R + P$, and we have $M^{-1} = R + P$. Choose $x \in M^{-1}$ such that $x \notin R$. Then $x = a/b$ where $a, b \in R$ and $b \neq 0$. Since $x \notin R$, we have $a \notin Rb$. Thus we can apply Theorem 82 and obtain

$$Rb = (Fa \cap Rb) + (Rb : a)a.$$

Therefore $b \in Fa$ and so $\frac{1}{x} = \frac{b}{a} \in F$. This shows that x is a unit in F. Since $N \cap M^{-1} = P$, it follows that x is a unit in M^{-1}. We have shown that $P \cap (M^{-1} - R)$ is empty. Hence we have $P \subset R$. But then $M^{-1} = R + P = R$ and this contradicts (4). Therefore M is a principal ideal of R.

(6) <u>Let R be a θ-ring with maximal ideal M and quotient field Q. Let S be an R-module such that $R \subset S \subset Q$ and $S^{-1} = 0$. Let x be any nonzero element of M. Then $Rx \neq Sx \cap R$.</u>

Suppose that $Rx = Sx \cap R$. Let b and c be nonzero elements of M. Suppose $b \notin Rc$, and let $a = xc \in Mc$. Then by Theorem 82

$$Rc = (Sa \cap Rc) + (Rc : b)b.$$

Now $Ra = Rxc = (Sx \cap R)c = Sxc \cap Rc = Sa \cap Rc$. Hence $Rc = Ra + (Rc : b)b \subset Ra + Rb \subset Mc + Rb$. Thus $Rc + Rb = Mc + Rb$, and hence $M(\frac{Rc + Rb}{Rb}) = \frac{Mc + Rb}{Rb} = \frac{Rc + Rb}{Rb}$. Therefore, by the Nakayama Lemma, we have $Rc + Rb = Rb$. Therefore, $c \in Rb$.

Thus if $b \notin Rc$, then $c \in Rb$, and this shows that R is a valuation ring. This is a contradiction since a valuation ring does not have a remote quotient field. Hence $Rx \neq Sx \cap R$.

(7) <u>θ-rings do not exist</u>

Let R be a θ-ring with maximal ideal M. By (5) there is an element $a \in R$, $a \neq 0$, such that $M = Ra$. Let F be the integral closure of R. Then we have $Ra = Fa \cap R$. However, since $F^{-1} = 0$ by (3), this contradicts (6). Therefore θ-rings do not exist.

By (2) a D-ring with a remote quotient field is either a θ-ring or a ring of type I. Since we have shown that θ-rings do not exist, R is a ring of type I. This concludes the proof of the theorem.

THEOREM 84. Let R be an integral domain whose quotient field Q is not remote. Let A be an extension of R by Q. Then A is decomposable if and only if R is a direct summand of A.

Proof. Suppose that $A = A_1 \oplus A_2$ is a nontrivial direct sum decomposition of A. Then A_1 and A_2 are torsion-free R-modules of rank 1. Since Q is not remote from R, Q is not the sum of two proper R-submodules. Hence if f is the canonical mapping of A onto Q with kernel R, then either $f(A_1) = Q$ or $f(A_2) = Q$. We can assume that $f(A_2) = Q$. Then we have $A = R + A_2$. Since A_2 has rank 1, and Q is torsion-free, we have $\text{Ker } f \cap A_2 = 0$. Therefore, $A = R \oplus A_2$.

The following theorem is an equivalent version of Theorem 83.

THEOREM 85. The following statements are equivalent:

(1) R is a ring of type I.

(2) R is a D-ring and is not complete (in the R-topology).

Proof. (1) $\Rightarrow$ (2). Let R be a ring of type I, and let M_1 and M_2 be the maximal ideals of R. Since R is an h-local ring, and R_{M_1} and R_{M_2} are maximal valuation rings, the completion of R (in the R-topology) is $R_{M_1} \oplus R_{M_2}$ by Theorem 22(5). Therefore, R is not complete in the R-topology. R is a D-ring by Theorem 78.

$(2) \Rightarrow (1)$.   Suppose that   R   is a D-ring and is not complete in the R-topology. If R has a remote quotient field, then R is a ring of type I by Theorem 83. Suppose that R does not have a remote quotient field. Let A be an extension of R by Q. Then A is a torsion-free R-module of rank 2, and hence is a direct sum of two modules of rank 1 by property D. By Theorem 84, A is a split extension of R by Q. Thus we have $\operatorname{Ext}_R^1(Q, R) = 0$, and thus by Theorem 9, R is complete in the R-topology. This contradiction shows that R is a ring of type I.

14

INTEGRALLY CLOSED D-RINGS

In Theorem 95 of this chapter we will finally solve the problem of finding all integrally closed D-rings. Theorems 86 through 89 prove some general facts about direct sum decompositions that will be needed at various stages of the proof. Theorems 90 and 91 are lemmas for Theorem 92, a theorem which enables us to drop down to a factor ring or go up to a localization. Theorems 93 and 94 are key lemmas for Theorem 95. Finally, Theorem 96 is a corollary of Theorem 95. The chapter closes with examples of D-rings.

THEOREM 86. Suppose that A is a torsion-free R-module of finite rank n, and that $A = A_1 + \ldots + A_m$, where $\sum\limits_{i=1}^{m} \text{rank } A_i \leq n$. Then $A = A_1 \oplus \ldots \oplus A_m$.

Proof. Let $B = A_1 \oplus \ldots \oplus A_m$; then the natural mapping $\emptyset: B \to A$ is surjective. Thus $\text{rank } B = \text{rank } A + \text{rank } (\text{Ker } \emptyset)$. But $\text{rank } B = \sum\limits_{i=1}^{m} \text{rank } A_i \leq \text{rank } A$, and hence $\text{rank } (\text{Ker } \emptyset) = 0$. It follows that $\text{Ker } \emptyset = 0$, and thus $\emptyset$ is an isomorphism.

THEOREM 87. Let R be a maximal valuation ring, and A a torsion-free R-module. Let B be a submodule of A such that A/B is a finite direct sum of k cyclic torsion R-modules, where k < rank A. Then there is a nonzero direct summand of B which is also a direct summand of A.

Proof. There exists a pure submodule F of A of rank $\leq$ k which maps onto A/B. Let C = F $\cap$ B; then C is a pure submodule of B. Hence by Theorem 51(2), C is a direct summand of B and there is a submodule D of B such that B = C $\oplus$ D.

We will prove that A = F $\oplus$ D. Now A = F + B = F + C + D = F + D. Let x $\in$ F $\cap$ D; since F/C is isomorphic to A/B, F/C is a torsion R-module. Hence there exists a nonzero element r in R such that rx $\in$ C. But then rx $\in$ C $\cap$ D = 0, and hence x = 0. Thus F $\cap$ D = 0, and we have A = F $\oplus$ D.

Since rank F < rank A, D is a nonzero module.

THEOREM 88. Let R be a ring, S and T R-modules, and D an injective submodule of S $\oplus$ T. Let E be an injective envelope of D $\cap$ S in D, and let F be a complementary summand of E in D. Thus D = E $\oplus$ F; and E and F project monomorphically into S and T, respectively.

Proof. The kernel of the projection of D into T is D $\cap$ S. Hence F projects monomorphically into T. Let f be the projection of E into S. Since Ker f $\subset$ T, we have Ker f $\cap$ (D $\cap$ S) = 0. However, E is an essential extension of D $\cap$ S, and thus Ker f = 0.

THEOREM 89. Let R be an integral domain and A a torsion-free, cotorsion R-module of finite rank. Suppose that A has two direct sum decompositions: A = B $\oplus$ C and A = $D_1 \oplus \ldots \oplus D_n$, where $\text{inj.dim}_R C = 1$ and B and $D_1, \ldots, D_n$ are indecomposable

R-modules. Then there exists an integer i, $1 \leq i \leq n$ such that $B \cong D_i$.

Proof. Let Q be the quotient field of R, and let $K = Q/R$. Let $K \otimes_R B = G$, $K \otimes_R C = E$, and $K \otimes_R D_i = H_i$ for $i = 1, \ldots, n$. Then we have $G \oplus E = H_1 \oplus \ldots \oplus H_n$. Since $\text{Hom}_R(K, G) \cong B$ and $\text{Hom}_R(K, H_i) \cong D_i$ by Theorem 6, it is necessary and sufficient to prove that $G \cong H_i$ for some $i = 1, \ldots, n$. By Theorem 7, G and H_i are indecomposable R-modules. Since $\text{inj.dim}_R C = 1$, E is an injective R-module.

Suppose that G is an injective R-module. By Theorem 88 $G = G_1 \oplus G_2$ where G_1 is isomorphic to a direct summand of H_1 and G_2 is isomorphic to a direct summand of $H_2 \oplus \ldots \oplus H_n$. Since G is indecomposable, either $G = G_1$ or $G = G_2$. If $G = G_1$, then $G \cong H_1$, since H_1 is indecomposable. Hence we can assume that $G = G_2$. Repeating this argument we see that either $G \cong H_2$ or G is isomorphic to a direct summand of $H_3 \oplus \ldots \oplus H_n$. Continuing in this way we find that G must be isomorphic to H_i for some $i = 1, \ldots, n$. Thus we can assume that G is not an injective R-module.

If every H_i is injective, then their direct sum is injective, and hence G, which is a direct summand of their direct sum, is injective. This contradiction shows that some of the H_i's are not injective. Let L_1 be the direct sum of the H_i's which are injective, and let L_2 be the direct sum of the H_i's which are not injective. Then $L_2 \neq 0$, and we have

$$E \oplus G = L_1 \oplus L_2 .$$

Now L_2 has no nonzero injective submodules. For suppose that M is a nonzero injective direct summand of L_2. Then $\text{Hom}_R(K, M)$ is a direct summand of $\text{Hom}_R(K, L_2)$, and thus $\text{Hom}_R(K, M)$ is a torsion-free cotorsion module of finite rank. Thus $\text{Hom}_R(K, M)$ is a direct sum of a finite number of indecomposable modules. Therefore, by the duality of Theorem 6, M is the direct sum of a finite number of inde-composable injective modules. Therefore, we can assume that M is an indecomposable injective module. By repeating the process carried out earlier with G we see that M is isomorphic to one of the H_i's making up L_2. This contradiction shows that L_2 has no nonzero injective submodules.

By Theorem 88, $E = E_1 \oplus E_2$, where E_1 projects mono-morphically into L_1 and E_2 projects monomorphically into L_2. Since L_2 has no nonzero injective submodules, $E = E_1$ maps monomorphically into L_1. Thus $L_1 = S_1 \oplus S_2$, where S_1 is the image of E and is iso-morphic to E. Thus there is a mapping f of $L_1 \oplus L_2$ onto S_1 with kernel $S_2 \oplus L_2$ which sends E isomorphically onto S_1. Hence we have $L_1 \oplus L_2 = E + \text{Ker} f$ and $E \cap \text{Ker} f = 0$. Thus $L_1 \oplus L_2 = E \oplus \text{Ker} f = E \oplus S_2 \oplus L_2$.

Therefore, we have $S_2 \oplus L_2 \cong \dfrac{L_1 \oplus L_2}{E}$. But $\dfrac{L_1 \oplus L_2}{E} \cong \dfrac{E \oplus G}{E} \cong G$. Hence we have proved that $G \cong S_2 \oplus L_2$. Since G is indecomposable and L_2 is a direct sum of H_i's, there is an H_i such that $G \cong H_i$.

THEOREM 90. Let P be a prime ideal of R such that $P = PR_P$ and R_P is a valuation ring. If A is a torsion-free R-module of finite rank n, then A/PA is a torsion-free R/P-module of rank $\leq n$. Conversely, if B is a torsion-free R/P-module of rank n, then there is a free R_P-module F of rank n and an R-submodule A of F such that $PF \subset A \subset F$ and $B \cong A/PF$.

Proof. Let A be a torsion-free R-module of rank n. Since $A/PA \subset A_P/PA$, it is sufficient to prove that A_P/PA is a torsion-free R/P-module of R/P rank $\leq$ n. Since $R_P PA_P = PR_P A = PA$, we see that $A_P/PA = A_P/R_P PA_P$. Thus A_P/PA is a module over R_P/PR_P, the quotient field of R/P, and hence A_P/PA is a torsion-free R/P-module. We will prove that rank $A_P/PA \leq n$ over R/P by induction on n, the rank of A over R. Suppose that rank $A = 1$. We may assume that $PA \neq A_P$, and consequently A_P is not the quotient field of R_P. Since R_P is a valuation ring, it follows that A_P is isomorphic to an ideal of R_P. Because the ideals of R_P are linearly ordered and $A_P \neq R_P PA_P$, it follows that A_P is isomorphic to a principal ideal of R_P. Therefore, A_P/PA is isomorphic to R_P/PR_P and hence has rank 1 over R/P.

We now assume that rank $A = n > 1$, and the assertion true for modules of rank n-1. Let C be a pure R_P-submodule of A_P of rank n-1, and let $D = A_P/C$. Then D is a torsion-free R_P-module of rank 1. Since D is a flat R_P-module we have an exact sequence:

$$0 \to C/PC \to A_P/PA_P \to D/PD \to 0.$$

By the case $n = 1$, D/PD is a torsion-free R/P-module of rank ≤ 1. By induction, C/PC is a torsion-free R/P-module of rank $\leq n-1$. Thus A_P/PA_P is a torsion-free R/P-module of rank $\leq n$.

Conversely, let B be a torsion-free R/P-module of rank n. Let F be a free R_P-module of rank n. Since R_P/P is the quotient field of R/P, F/PF is a direct sum of n copies of the quotient field of R/P. Thus we can assume that B is an R/P-submodule (and hence an R-submodule) of F/PF. Thus there is an R-submodule A of F such that $PF \subset A \subset F$ and $B \cong A/PF$.

THEOREM 91. Let P be a prime ideal of R such that $P = PR_P$ and R_P is a maximal valuation ring. Then R is a D-ring if and only if R/P is a D-ring.

Proof. Suppose that R is a D-ring. Let B be a torsion-free R/P-module of rank $n > 1$. By Theorem 90 there is a free R_P-module F of rank n and an R-submodule A of F such that $PF \subset A \subset F$ and $A/PF \cong B$. Since R is a D-ring, $A = A_1 \oplus \ldots \oplus A_n$, where A_i has rank 1. Thus $A/PA \cong A_1/PA_1 \oplus \ldots \oplus A_n/PA_n$. By Theorem 90, A_i/PA_i is a torsion-free R/P-module of rank ≤ 1.

Since $PA \subset PF$, we have an R/P-homomorphism of A/PA onto B. Thus $B = B_1 + \ldots + B_n$, where B_i is the image of A_i/PA_i and hence is a torsion-free R/P-module of rank ≤ 1. By Theorem 86, $B = B_1 \oplus \ldots \oplus B_n$. Therefore R/P is a D-ring.

Conversely, assume that R/P is a D-ring. We will assume that R is not a D-ring and arrive at a contradiction. Let n be the smallest integer greater than 1 for which there is an indecomposable

torsion-free R-module of rank n, and let A be an indecomposable
torsion-free R-module of rank n. Then A is a reduced R-module.

By Theorem 90, A/PA is a torsion-free R/P-module whose
R/P rank is less than or equal to n. Since PA is an R_P-module, and
R_P has property D, PA is a direct sum of n R_P-modules of rank 1.
Hence we see that $A \neq PA$.

Since R/P has property D, A/PA is a finite direct sum of
torsion-free R/P-modules of rank 1. Therefore,
$A/PA = B_1/PA \oplus B_2/PA$, where B_1, B_2 are submodules of A such
that $A = B_1 + B_2$, $B_1 \cap B_2 = PA$, and rank $B_i/PA = k_i < n$ for $i = 1, 2$.

We will prove that B_1 and B_2 are each direct sums of n
R-modules of rank 1. For this we may assume that $B_1 \neq PA$ and
$B_2 \neq PA$, since PA is such a direct sum. Let $C = B_1/PA$; then
$C = C_1 \oplus \ldots \oplus C_{k_1}$ where C_j is a torsion-free R/P-module of rank 1
and $k_1 < n$. By Theorem 90, $C_{j_P} \cong R_P/P$. Therefore,
$C_P = C_{1_P} \oplus \ldots \oplus C_{k_{1_P}}$ is the direct sum of k_1 cyclic R_P-modules.
But $C_P = B_{1_P}/PA$, and B_1 is a torsion-free R_P-module of rank n.
Thus we can apply Theorem 87 and find a direct sum decomposition
$B_{1_P} = L \oplus M$, where $L \neq 0$ and $L \subset PA$. It follows that
$B_1 = L \oplus (M \cap B_1)$.

Since $B_1 \neq PA$, we see that $M \cap B_1 \neq 0$. But then L and
$M \cap B_1$ are nonzero torsion-free R-modules each of rank $< n$. By the
minimality of the integer n both L and $M \cap B_1$ are direct sums of
R-modules of rank 1. Hence in this case also B_1 is a direct sum of
n modules of rank 1. Similarly, B_2 is a direct sum of n modules of
rank 1.

Since $PA = B_1 \cap B_2$ and $B_1 + B_2 = A$, we have an exact sequence

$$0 \to PA \to B_1 \oplus B_2 \to A \to 0.$$

Since R_P is a maximal valuation ring and PA is a torsion-free R_P-module, we have by Theorem 51 that $\text{inj.dim}_{R_P} PA = 1$. Hence by Theorem A3, we have $\text{inj.dim}_R PA \leq \text{inj.dim}_{R_P} PA = 1$.

Using Theorem A3 and Theorem 51 we have $\text{Ext}^1_R(Q \otimes_R A, PA) \cong \text{Ext}^1_{R_P}(Q \otimes_R A, PA) = 0$. Since $\text{inj.dim}_R PA = 1$, we have $\text{Ext}^2_R(K \otimes_R A, PA) = 0$. From this it follows that $\text{Ext}^1_R(A, PA) = 0$. This proves that the preceding exact sequence splits, and we have

$$PA \oplus A \cong B_1 \oplus B_2 \ .$$

By the preceding paragraph we see that $B_1 \oplus B_2 = D_1 \oplus \ldots \oplus D_{2n}$, where each D_i is a reduced torsion-free R-module of rank 1.

As we have just seen, $\text{Ext}^1_R(Q, PA) = 0$. Therefore PA is a cotorsion R-module. Since A/PA is a torsion module of bounded order, we have $\text{Ext}^1_R(Q, A/PA) = 0$. From these facts it follows that $\text{Ext}^1_R(Q, A) = 0$. Thus A is also a cotorsion R-module. Hence $PA \oplus A$ is a cotorsion R-module. We can now apply Theorem 89 and conclude that $A \cong D_1$ for some i, $1 \leq i \leq 2n$. Thus A has rank 1. This contradiction shows that R is a D-ring.

THEOREM 92. Let R be an integrally closed ring and P a prime ideal of R such that $P = PR_P$. Then R is a D-ring if and only if both R_P and R/P are D-rings.

Proof. Assume that R is a D-ring. Then R_P is a D-ring by Theorem 59. Since R_P is integrally closed, R_P is a Prufer ring by Theorem 68. Therefore, R_P is a valuation ring. By Theorem 65, R_P is a maximal valuation ring. Hence by Theorem 91, R/P is also a D-ring.

Conversely, assume that both R/P and R_P are D-rings. As in the preceding paragraph R_P is a maximal valuation ring. Therefore, R is a D-ring by Theorem 91.

THEOREM 93. Let R be a valuation ring and P a prime ideal of R. Then R is a maximal valuation ring if and only if both R_P and R/P are maximal valuation rings.

Proof. By Theorem 65, a valuation ring is a D-ring if and only if it is a maximal valuation ring. Thus Theorem 93 is an immediate consequence of Theorem 91.

This is certainly a very roundabout way of proving this theorem and there is a relatively direct homological proof involving theorems on change of rings, but there seems to be no point in introducing that proof here.

THEOREM 94. Let R be a Prufer ring which is not a valuation ring and whose quotient field Q is not remote, and let J be the Jacobson radical of R. If M is a maximal ideal of R, then $R_M^{-1} = P$ is a nonzero prime ideal of R contained in J and P contains every prime ideal of R that is contained in J. We have $P^{-1} = R_P$ and $PR_P = P$. If N is any other maximal ideal of R, then $R_N^{-1} = P$ also.

Proof. Since Q is not remote from R, we have $P \neq 0$. Clearly, P is a proper ideal of R_M, and thus $P \subset MR_M \cap R = M$. Let N be another maximal ideal of R, and suppose that $P \not\subset N$. Then

$R = P + N$, and hence $1 = a + b$, where $a \epsilon P$ and $b \epsilon N$. Since $a \epsilon M$, we have $b \epsilon R - M$. Therefore $1/b \epsilon R_M$, and $a/b \epsilon PR_M \subset R$. Thus $1/b = a/b + 1 \epsilon R$, and this contradicts $b \epsilon N$. This contradiction shows that $P \subset N$. We have proved that $P \subset J$.

Let x be an element of R_M such that $x^n \epsilon P$ for some integer $n > 0$. If $y \epsilon R_M$, then $(xy)^n = x^n y^n \epsilon PR_M \subset R$, and xy is integral over R. But R is integrally closed, and thus $xy \epsilon R$. Hence we have $xR_M \subset R$, and this shows that $x \epsilon P$. Thus P is a radical ideal of R_M. Since R_M is a valuation ring, P is a prime ideal of R_M. A fortiori, P is a prime ideal of R.

Since $P \subset M$, we have $R_M \subset R_P$. Thus R_P is a valuation ring and $PR_P \subset MR_M$. Hence PR_P is a prime ideal of R_M. Now $P \cap R = P = PR_P \cap R$; therefore by the one-to-one correspondence between the prime ideals of R_M and the prime ideals of R contained in M, we have $P = PR_P$. Thus $P \subset R_P^{-1}$; but R_P^{-1} is a proper ideal of R_P, and hence $R_P^{-1} \subset PR_P = P$. Thus we see that $P = R_P^{-1}$.

Now $(P^{-1}P)R_M = P^{-1}(PR_M) = P^{-1}P \subset R$. Thus $P^{-1}P \subset R_M^{-1} = P$. This shows that P^{-1} is a ring. Since $R_P \subset P^{-1}$, P^{-1} is a valuation ring with maximal ideal $m(P^{-1}) \subset PR_P = P$. But P is a proper ideal of P^{-1}, and thus $P = m(P^{-1})$. Therefore, we have $P^{-1} = R_P$.

Let P' be a prime ideal of R contained in J. Then $R_N \subset R_{P'}$, and hence $P'R_{P'} \subset R_N$ for every maximal ideal N of R. Thus $P'R_{P'} \subset \bigcap_N R_N = R$, and $P' \subset R_{P'}^{-1} \subset R_M^{-1} = P$. Therefore, P contains every prime ideal of R that is contained in J. From this it follows immediately that $P = R_N^{-1}$ for every maximal ideal N of R.

We are now ready to prove the main theorem of these notes.

THEOREM 95. Let R be an integrally closed ring. Then R is a D-ring if and only if R is the intersection of at most two maximal valuation rings; i.e., if and only if R has only two maximal ideals M_1, M_2 (not necessarily distinct) and R_{M_1}, R_{M_2} are maximal valuation rings.

Proof. Assume that R is a D-ring. By Theorem 68, R is a Prüfer ring. If M is a maximal ideal of R, then R_M is a valuation ring, and hence by Theorem 65, R_M is a maximal valuation ring. Since $R = \bigcap_M R_M$, where M ranges over all maximal ideals of R, we must prove that R has at most two maximal ideals. We will suppose that R has more than two maximal ideals, and arrive at a contradiction.

Since R is not a ring of type I, we have by Theorem 83 that R does not have a remote quotient field. Thus by Theorem 94, there is a nonzero prime ideal P contained in the Jacobson radical J of R such that $PR_P = P$. Let $\overline{R} = R/P$; then by Theorem 91, $\overline{R}$ is a D-ring.

Let $\overline{M}$ be a maximal ideal of $\overline{R}$. We have $\overline{M} = M/P$, where M is a maximal ideal of R. Then $\overline{R}_{\overline{M}} \cong R_M/PR_M$ is a valuation ring, and thus $\overline{R}$ is a Prüfer ring. Since $P \subset J$, $\overline{R}$ has the same number of maximal ideals as R. Therefore, $\overline{R}$ is not a ring of type I, and thus $\overline{R}$ does not have a remote quotient field by Theorem 83. Hence we can apply Theorem 94 again and obtain a nonzero prime ideal P^* of $\overline{R}$ contained in the radical $\overline{J}$ of $\overline{R}$. Now $\overline{J} = J/P$; and

$P^* = P'/P$, where P' is a prime ideal of R such that $P \subsetneq P' \subset J$. But this contradicts Theorem 94 which asserts that P contains every prime ideal of R contained in J. Therefore, R is the intersection of at most two maximal valuation rings.

Conversely, assume that R is the intersection of at most two maximal valuation rings. If R is a maximal valuation ring or a ring of type I, then R has property D either by Theorem 65 or by Theorem 83. Thus we can assume that R is the intersection of two independent maximal valuation rings and is not a ring of type I.

Let M_1 and M_2 be the two maximal ideals of R. Then R_{M_1} and R_{M_2} are maximal valuation rings by Theorem 52. Since R is not a ring of type I, there is a nonzero prime ideal P of R such that $P \subset M_1 \cap M_2$. Then $R_{M_1} \subset R_P$, and hence R_P is a maximal valuation ring by Theorem 93. We have $PR_P \subset R_{M_1} \cap R_{M_2} = R$, and thus $PR_P = P$.

By Zorn's Lemma (or by Theorem 94) we can assume that P contains every prime ideal of R contained in $M_1 \cap M_2$. Thus $\overline{R} = R/P$ has exactly two maximal ideals $\overline{M}_1 = M_1/P$ and $\overline{M}_2 = M_2/P$; and $\overline{M}_1 \cap \overline{M}_2$ contains no nonzero prime ideal of R. By Theorem 93, $\overline{R}_{\overline{M}_i} \cong R_{M_i}/P_{M_i}$ is a maximal valuation ring for $i = 1, 2$. Therefore, $\overline{R}$ is a ring of type I; and hence $\overline{R}$ is a D-ring by Theorem 83. Thus R is a D-ring by Theorem 91.

THEOREM 96. Let R be a D-ring. Then R has at most two maximal ideals.

Proof. Let F be the integral closure of R. Then F is a D-ring by Theorem 59. Hence by Theorem 95, F has at most two maximal ideals. Therefore, R has at most two maximal ideals.

Remark: We note that it follows easily from Theorem 95 that if R is a D-ring, then its lattice of prime ideals can be represented symbolically by $|$, $||$, or by Y .

Examples:

(1) The first example is of nonNoetherian quasi-local D-ring which is not a maximal valuation ring. It is easily seen that this ring is isomorphic to the example given in Chapter 11. However, the constructtion is different and presents this example in a new light. We will use Theorem 91 to prove that it is a D-ring.

Let k be a field and X and Y indeterminates over k. Let A be the ring of formal power series in Y with coefficients in k and nonnegative integer exponents: $A = k[[Y]]$. Let B be the quotient field of A, and let F be the ring of formal power series in X with coefficients in B and nonnegative integer exponents, but with constant term in A:

$$F = \{ \sum_{i=0}^{\infty} b_i X^i \mid b_0 \in A,\ b_i \in B \text{ for } i > 0\}.$$

It is easily seen that F is a valuation ring. Let P be the prime ideal of F consisting of power series with constant term $b_0 = 0$. Then $F_P = B[[X]]$ is a complete discrete valuation ring and $F/P \cong A = k[[Y]]$ is also a complete discrete valuation ring. Therefore by Theorem 93, F is a maximal valuation ring. F is not a Noetherian ring, since it is a rank 2 valuation ring.

Let A' be the subring of A consisting of power series in Y with linear term missing. Then A' is a complete Noetherian local ring of Krull dimension 1 such that every ideal of A' can be generated by two elements. Thus A' is a D-ring by Theorem 64.

Let R be the subring of F of power series in X with constant term $b_0 \epsilon A'$. Let P be the same prime ideal as in F. Then $R_P = F_P = B[[X]]$ is a complete discrete valuation ring; $R/P \cong A'$ is a D-ring; and $PR_P = P$. Thus by Theorem 91, R is a D-ring. R is a quasi-local ring with maximal ideal consisting of those power series in X with constant term in the maximal ideal of A'. F can be generated over R by two elements, and hence R is not a Noetherian ring, since F is not Noetherian.

(2) The second example is of a ring which is the intersection of two independent maximal valuation rings, but is not a ring of type I. Mrs. Osofsky has communicated to me an example of this type of ring, but our example illustrates more easily the methods of these notes. Together with the example of Barbara Osofsky of a ring of type I given in Chapter 12, this example proves the existence of rings which are intersections of two maximal valuation rings whether of type I or not.

Let A be a ring of type I (in particular, A could be the example of Barbara Osofsky). Let B be the quotient field of A, and let X be an indeterminate over B. Let R be the ring of formal power series in X with coefficients in B and nonnegative integer exponents, but with constant term in A:

$$R = \{\sum_{i=0}^{\infty} b_i X^i \mid b_0 \epsilon A, \ b_i \epsilon B \ \text{for} \ i > 0\}.$$

Let P be the prime ideal of R consisting of power series with constant term $b_0 = 0$. Then $R/P \cong A$ is a D-ring by Theorem 83; and $R_P \cong B[[X]]$ is a complete discrete valuation ring. Clearly $PR_P = P$. Thus R is a D-ring by Theorem 91.

Let M_1 and M_2 be the two maximal ideals of A. Then $N_1 = M_1 + P$ and $N_2 = M_2 + P$ are the only two maximal ideals of R, and $P \subset N_1 \cap N_2$. Thus R is not a maximal valuation ring or a ring of type I.

R_{N_1} is the ring of power series in X with constant term $b_0 \in A_{M_1}$. Clearly R_{N_1} is a valuation ring, since A_{M_1} is a valuation ring. $(R_{N_1})_P \cong B[[X]]$ is a complete discrete valuation ring; and $R_{N_1}/P \cong A_{M_1}$ is a maximal valuation ring. Thus by Corollary 2, R_{N_1} is a maximal valuation ring. Similarly, R_{N_2} is a maximal valuation ring. Thus $R = R_{N_1} \cap R_{N_2}$ is the intersection of two independent maximal valuation rings, and R is not a ring of type I.

15

HAUSDORFF D-RINGS

In Theorem 64 of Chapter 10 we proved that a domain is a Noetherian D-ring if and only if it is a ring of type II. The main burden of this chapter is to show (as we will in Theorem 100) that Noetherian D-rings are precisely the D-rings that satisfy the Hausdorff conditions that $\bigcap_n I^n = 0$ for every ideal I of R. We thus find that the question of whether or not a D-ring is Noetherian is really a topological question.

THEOREM 97. Let R be an integral domain whose quotient field Q is not remote $(Q \neq R)$, and suppose that $\bigcap_n I^n = 0$ for every proper principal ideal I of R. Then R is a quasi-local ring of Krull dimension 1.

Proof. Suppose that R has two distinct nonzero prime ideals P_1 and P_2. We can assume that $P_1 \not\subset P_2$. Choose an element $a \in P_1$ such that $a \notin P_2$, and let $S = \{a^n\}$ be the multiplicatively closed set generated by a. Now $R_S^{-1} = \bigcap_n Ra^n$, and therefore $R_S^{-1} = 0$ by assumption. Since Q is not remote from R we conclude that $R_S = Q$. However, $P_2 \cap S$ is empty, and thus $R_S P_2$ is a nonzero proper, prime ideal of R_S. Therefore, R_S cannot be a field. This contradiction shows that R has only one nonzero prime ideal.

THEOREM 98. Let R be a D-ring whose quotient field Q is not remote $(Q \neq R)$, and suppose that $\bigcap_n I^n = 0$ for every proper, principal ideal I of R. Then the integral closure of R is a maximal valuation ring of Krull dimension 1.

Proof. Let F be the integral closure of R. By Theorem 97, R is a quasi-local ring of Krull dimension 1. Thus F also has Krull dimension 1. By Theorem 72, F is a maximal valuation ring.

THEOREM 99. Let R be an integrally closed domain. Then the following statements are equivalent:

(1) R is a D-ring and $\bigcap_n I^n = 0$ for every proper, principal ideal I of R.

(2) R has Krull dimension 1, and R is either a maximal valuation ring or a ring of type I.

Proof. (1) $\Rightarrow$ (2). Suppose that R is a D-ring and $\bigcap_n I^n = 0$ for every proper, principal ideal of R. If R has a remote quotient field, then R is a ring of type I by Theorem 83. If R does not have a remote quotient field, then R is a maximal valuation ring by Theorem 98. Thus R is either a maximal valuation ring or a ring of type I. We will prove that R has Krull dimension 1.

Suppose that R does not have Krull dimension 1. Then R has a nonzero prime ideal P which is not a maximal ideal. Suppose that R is a valuation ring. Then there exists a non-unit $a \in R$ such that $a \notin P$. It follows that $P \subset Ra^n = 0$ for all n. Thus $P \subset \bigcap Rb^n = 0$. This contradiction shows that R is a ring of type I.

Let M_1 and M_2 be the two maximal ideals of R. Since R is an h-local ring, we can assume that $P \subsetneq M_1$, and $P \not\subset M_2$. If $(M_1 - P) \subset M_2$, then $P \subset M_2$. Hence there exists an element $b \in M_1$ such that $b \notin P$ and $b \notin M_2$.

Now if I is any ideal of R that is contained in M_1, but not in M_2, then $R_{M_1} I \cap R = I$ by Theorem 25. Thus for any integer $n > 0$, we have $R_{M_1} b^n \cap R = Rb^n$ and $R_{M_1} P \cap R = P$. Since $b^n \notin P$ and R_{M_1} is a valuation ring, we have $R_{M_1} P \subset R_{M_1} b^n$. Thus $P = (R_{M_1} P \cap R) \subset (R_{M_1} b^n \cap R) = Rb^n$ for all $n > 0$. Therefore $P \subset \bigcap Rb^n = 0$. This contradiction shows that R has Krull dimension 1.

(2) $\Rightarrow$ (1). Assume that R is either a maximal valuation ring or a ring of type I. By Theorem 65 or by Theorem 83, R is a D-ring. We now assume that R also has Krull dimension 1, and we must show that $\bigcap I^n = 0$ for every proper principal ideal I of R.

Let r be any non-zero element of the Jacobson radical of R, and let $S = \{r^n\}$, the multiplicatively closed subset generated by r. Since every non-zero prime ideal of R meets S, we see that R_S is the quotient field Q of R. Let $J = \bigcap Rr^n$, and suppose $J \neq 0$. Take $a \in J$ ($a \neq 0$). Since $R_S = Q$, there exists $b \in R$ and an integer $n > 0$ such that $1/a = b/r^n$. But $a = cr^{n+1}$ for some $c \in R$, and hence $cbr = 1$. Thus r is a unit in R, and this is a contradiction. Therefore $\bigcap Rr^n = 0$.

This disposes of the case where R is a valuation ring, and hence we may assume that R is a ring of type I with two maximal ideals M_1 and M_2. In the light of the preceding paragraph it will be sufficient to show that if $b \in M_1$ and $b \notin M_2$, then $\bigcap Rb^n = 0$. Now

R_{M_1} is a quasi-local ring of Krull dimension 1, and hence by the preceding paragraph, $\bigcap_n R_{M_1} b^n = 0$. It follows from Theorem 25 that $R_{M_1} b^n \cap R = Rb^n$. Therefore $\bigcap_n Rb^n = 0$.

THEOREM 100. Let R be a D-ring. Then the following statements are equivalent:

(1) R is a Noetherian ring.

(2) $\bigcap_n I^n = 0$ for every ideal I of R.

Either of these conditions is equivalent to R being a ring of type II.

Proof. If R is a Noetherian domain, it is an elementary fact that $\bigcap_n I^n = 0$ for every ideal I of R. Conversely, assume that $\bigcap_n I^n = 0$ for every ideal I of R. We will prove that R is a Noetherian ring. The equivalence of the condition that R is a Noetherian D-ring with the condition that R is a ring of type II was established in Theorem 64.

We will first show that R does not have a remote quotient field. Suppose that R does have a remote quotient field. Then by Theorem 83, R is a ring of type I. Let M_1 and M_2 be the maximal ideals of R. Then we have

$$\bigcap_n (R_{M_1} M_1)^n = \bigcap_n (R_{M_1} M_1^n) = R_{M_1} \left(\bigcap_n M_1^n \right) = 0.$$

Since R_{M_1} is a maximal valuation ring, this implies that R_{M_1} is a complete discrete valuation ring. Similarly R_{M_2} is a complete discrete valuation ring. Since R is an h-local ring, we have by Theorem 26 that R is a Noetherian ring. But Noetherian rings of type I do not exist, as a consequence of Theorem 64. This contradiction shows that R does not have a remote quotient field.

We now have by Theorem 97 that R is a quasi-local ring of Krull dimension 1 with maximal ideal M. Let F be the integral closure of R. Then by Theorem 98, F is a maximal valuation ring of Krull dimension 1 with maximal ideal N. We assert that N is a principal ideal of F.

Suppose that N is not a principal ideal of F. Now by Theorem 56 we have $\dim_{R/M} F/FM \leq 2$. If $FM \neq N$, then there exists an element $x \in N - FM$ and $FM \subsetneq Fx \subset N$. But $\dim_{R/M} N/FM = 1$ in this case, and thus $N = Fx$ is a principal ideal of F. Thus we can assume that $FM = N$. Let $I = F^{-1}$. Since R does not have a remote quotient field, we have $I \neq 0$. Since I is an ideal of F, we have for every integer $k > 0$ that $IN^k = I(FM)^k = I(FM^k) = IM^k \subset M^k$. Therefore, $I(\bigcap_k N^k) \subset \bigcap_k IN^k \subset \bigcap_k M^k = 0$. Hence $\bigcap N^k = 0$. But since F is a valuation ring, this implies that N is a principal ideal of F. Thus in all cases we have that N is a principal ideal of F; and since F has Krull dimension 1, we see that $\bigcap N^k = 0$. From this it follows immediately that F is a complete discrete valuation ring.

Let x be an element of F such that $N = Fx$. Every ideal of F is a power of N. Now $\dim_{R/M} F/N \leq \dim_{R/M} F/FM \leq 2$. Since $Fx^i/Fx^{i+1} \cong F/Fx = F/N$; it follows that if J is any nonzero ideal of F, then F/J is an R-module of finite length. Let $I = F^{-1}$; since R does not have a remote quotient field, I is a nonzero ideal of F that is contained in R. We have just seen that F/I is an R-module of finite length. Since F/I maps onto F/R, we see that F/R is an R-module of finite length. Thus F is a finitely generated R-module.

Since every ideal of F is isomorphic to F, every ideal of F is a finitely generated R-module. Thus I is a finitely generated ideal of R. Now M/I is an R-submodule of F/I. Thus M/I is an R-module of finite length. Therefore, M is a finitely generated ideal of R. Since M is the only nonzero prime ideal of R, it follows from Theorem 49 that R is a Noetherian ring.

CONCLUSION

We have now completely described the present state of our knowledge concerning the theory of D-rings. However, there are a number of questions that remain unanswered. We have not yet found necessary and sufficient conditions for a quasi-local domain to be a D-ring. We know that a quasi-local D-ring R is complete in the R-topology, that it does not have a remote quotient field, and that its integral closure is a maximal valuation ring. We conclude from this that its lattice of prime ideals is linearly ordered. But we know by examples that R need not be a maximal valuation ring. Theorem 75 provides us with sufficient conditions for a quasi-local domain to be a D-ring, but we don't know whether or not they are necessary. In fact, it is doubtful if they are. Theorem 71 shows that there is a completely reflexive ring between a quasi-local D-ring R and its integral closure, if the D-ring is not integrally closed; but we don't know whether or not R must be a completely reflexive ring. All of our examples are of this kind, but again it is doubtful if this is a necessary condition.

If a D-ring R has more than one maximal ideal, then we know that it can only have two maximal ideals. If R is h-local, then it is a ring of type I. We don't know, however, whether there exist any

162

D-rings with more than one maximal ideal that are not integrally closed; that is, that are not the intersection of two maximal valuation rings. We do know, however, that if such a ring R exists, then it is complete in the R-topology, does not have a remote quotient field, and that its lattice of prime ideals can be described by the symbol Y .

In addition to the problems suggested here, there are many other unsolved questions outstanding in the theory of torsion-free modules. Perhaps the tools and methods presented here will be most useful if they can be applied to, and stimulate further research in, the general theory of torsion-free modules over a domain.

Bibliography

1. Bass, H.   Torsion-free and projective modules. Trans. Amer.
   Math. Soc. 102 (1962): 319-27.

2. _____ . On the ubiquity of Gorenstein rings. Math. Z. 82 (1963):
   8-28.

3. Bourbaki, N. Eléments de mathématique: algèbre commutative,
   fasc. 27, no. 1920. Paris: Hermann, 1961.

4. _____ . Eléments de mathématique: algèbre commutative, fasc. 31,
   no. 1314. Paris: Hermann, 1965.

5. Cartan, H., and Eilenberg, S. Homological algebra. Princeton
   University Press, 1956.

6. Cohen, I. S. Commutative rings with restricted minimum
   condition. Duke Math. J. 17 (1950): 27-42.

7. Eckmann, B., and Schopf, A. Über injective Moduln.
   Arch. Math. 4 (1953): 75-78.

8. Fleischer, I.   Modules of finite rank over Prüfer rings.
   Annals of Math. 65, no. 2 (1957): 250-54.

9. Harrison, D.   Infinite Abelian groups and homological methods.
   Annals of Math. 69, no. 2 (1959): 366-91.

10. Hattori, A. On Prüfer rings. J. Math. Soc. Japan 9, no. 4 , (1957): 381-85.

11. Jans, J. Duality in Noetherian rings. Proc. Amer. Math. Soc. 12 (1961): 829-35.

12. Kaplansky, I. Modules over Dedekind rings and valuation rings. Trans. Amer. Math. Soc. 72 (1962): 327-40.

13. _____. Dual modules over a valuation ring I. Proc. Amer. Math. Soc. 4, no. 2 (1953): 213-17.

14. _____. Decomposability of modules. Proc. Amer. Math. Soc. 13 (1962): 532-35.

15. _____. Infinite Abelian groups. Ann Arbor: University of Michigan Press, 1969.

16. Matlis, E. Injective modules over Noetherian rings. Pac. J. Math. 8, no. 3 (1958): 511-28.

17. _____. Injective modules over Prufer rings. Nagoya Math. J. 15 (1959): 57-69.

18. _____. Some properties of Noetherian domains of dimension 1. Can. J. Math. 3 (1959): 222-41.

19. _____. Divisible modules. Proc. Amer. Math. Soc. 11, no. 3 (1960): 385-91.

20. _____. Cotorsion modules. Mem. Amer. Math. Soc., no. 49, 1964.

21. _____. Decomposable modules. Trans. Amer. Math. Soc. 125 (1966): 147-79.

22. Matlis, E.   The decomposability of torsion-free modules of
    finite rank. Trans. Amer. Math. Soc. 134, no. 2 (1968):
    315-24.

23. _____.   Reflexive domains. J. Algebra 8, no. 1 (1968): 1-33.

24. _____.   The two-generator problem for ideals. Mich. Math. J.
    17 (1970): 257-65.

25. _____.   Rings of type I. J. Algebra 23(1972).

26. _____.   Rings of Type II. Mich. Math. J. 19(1972): 141-47.

27. _____.   Local D-rings. Math. Z. 124(1972): 266-72 .

28. _____.   Rings with property D. Trans. Amer. Math. Soc.
    170, no. 2, (1972) .

29. Nagata, M. Local rings . New York: Interscience Publ., 1962.

30. Northcott, D. Ideal Theory. London: Cambridge University
    Press, 1953.

31. Nunke, R.   Modules of extensions over Dedekind rings.
    Illinois J. Math. 3, no. 2 (1959): 222-41.

32. Prufer, H.   Theorie der abelschen Gruppen, II: Ideale Gruppen.
    Math. Z. 22 (1925): 222-49.

33. Schilling, O. The theory of valuations. Math. Surveys, no. 4.
    Amer. Math. Soc., 1950.

34. Schmidt, F.   Mehrfach perfekte Körper. Math. Ann. 108 (1933):
    1-25.

35.   Zariski, O., and Samuel, P.   Commutative Algebra, Vols. I and
      and II.   Princeton, N. J.: Van Nostrand, 1960.

36.   Zelinsky, D.   Complete fields from local rings.   Proc. Nat.
      Acad. Sciences. 37, no. 6 (1951): 379-81.

37.   _____.   Linearly compact modules and rings.   Amer. J. Math.
      75, no. 1 (1953): 79-90.

INDEX